GLOBAL CLIMATE CHANGE

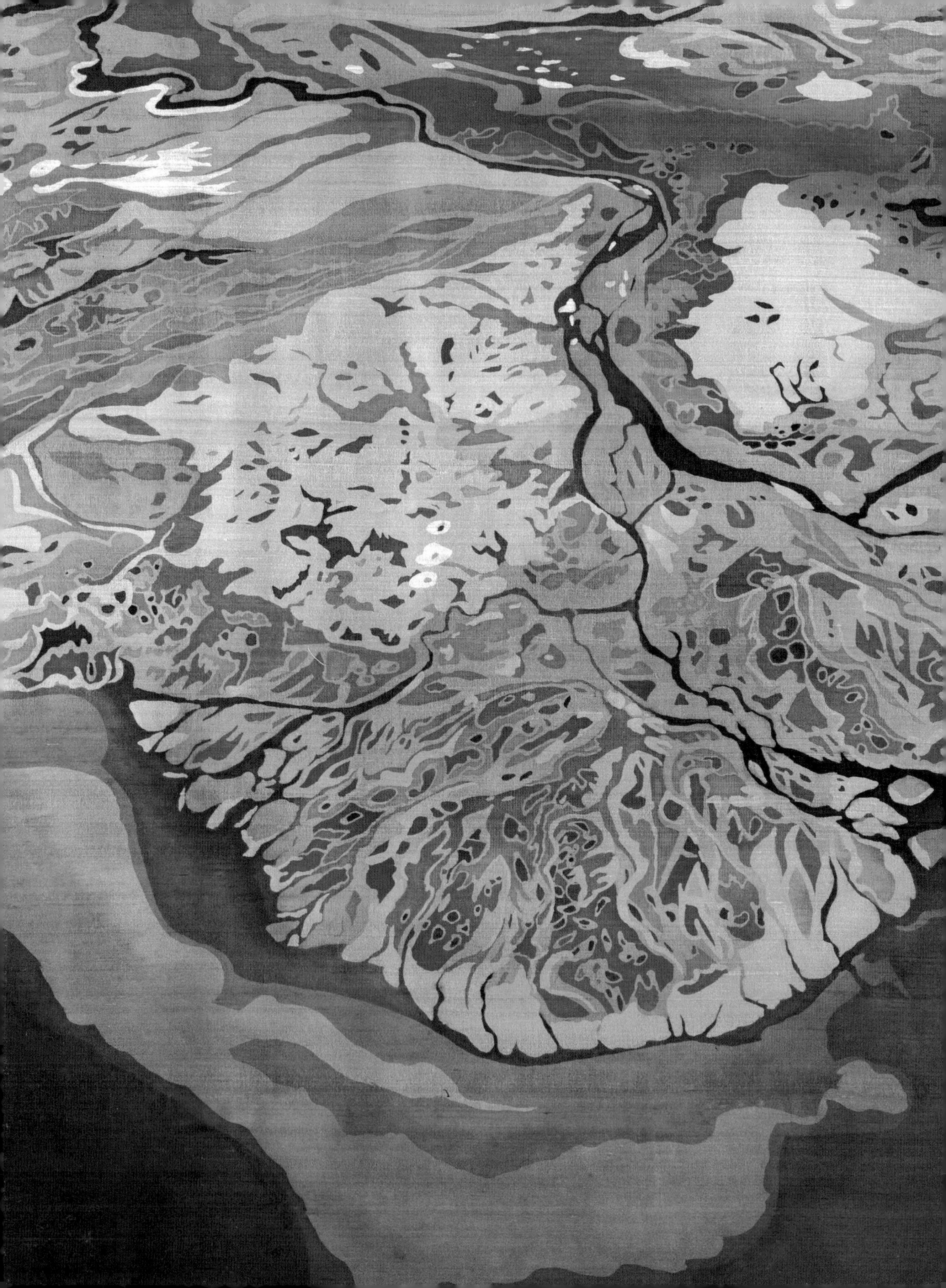

GLOBAL CLIMATE CHANGE

ORRIN H. PILKEY AND KEITH C. PILKEY

WITH BATIK ART BY MARY EDNA FRASER

DUKE UNIVERSITY PRESS DURHAM AND LONDON 2011

Printed in Italy on acid-free paper ∞
Designed by C. H. Westmoreland
Typeset in Chaparral with Gill Sans display
by Copperline Book Services, Inc.

Library of Congress Cataloging-in-Publication Data
appear on the last printed page of this book.

DUKE UNIVERSITY PRESS GRATEFULLY ACKNOWLEDGES THE SUPPORT OF THE FOLLOWING ORGANIZATIONS AND INDIVIDUALS FOR PROVIDING FUNDS TOWARD THE PRODUCTION OF THIS BOOK.

The Santa Aguila Foundation (www.coastalcare.org) is a nonprofit organization dedicated to preserving coastlines around the world. Global climate change is central to the future of the world's coasts; that is especially true of rising sea level, which amplifies coastal erosion, the melting of ice caps and glaciers, and the destruction of reefs, wetlands, and ecosystems. More intense rains and hurricanes combining with rising sea levels will lead to more severe flooding and potential loss of property and life. The problem is serious. More than 10% of the world's population lives in vulnerable areas less than ten meters (about thirty feet) above sea level. Eleven of the fifteen largest cities in the world are located on a coast or river estuary. The rising sea level will potentially result in millions of climate refugees. It is already a threat to many island nations that are at the front lines of climate change. Currently many people are skeptical about the very existence of global climate change, let alone its human connection. We believe that there can be no meaningful response to this global challenge without public acceptance of the scientific facts. This is why the Santa Aguila Foundation supports this book. The more knowledgeable and educated we become about our shorelines and the threats they face from global climate change, the more connected we become, and the firmer our willingness to protect these endangered natural habitats will be. These conservation efforts will have an impact on all living beings, but it is our children and future generations who will benefit the most. Ultimately, global climate change will affect life on Earth in many ways, but the extent of the change is largely up to us. We hope that readers will understand the need for immediate action if the world's beaches and coastlines are to remain beautiful playgrounds and important ecosystems.

The mission of the Coastal Conservation League (coastalconservationleague.org) is to protect the natural environment of the South Carolina coastal plain and to enhance the quality of life of our communities.

The North Carolina Coastal Federation (nccoast.org) provides citizens and groups with the assistance they need to take an active role in the stewardship of North Carolina's coastal water quality and natural resources.

WE DEDICATE THIS BOOK TO

All of our families

Melanie Taylor, a profile in courage

Dana H. Dixon, a dedicated conservationist

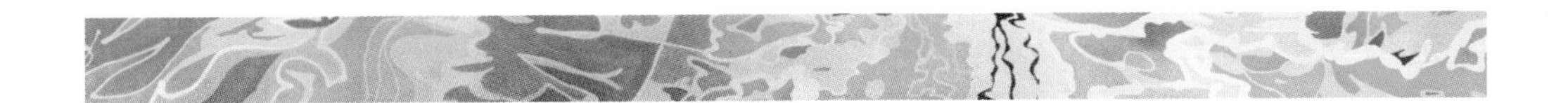

CONTENTS

PREFACE

GLOBAL CHANGE IS UPON US. Of this there is no doubt among those who observe the Earth. But there are two separate issues. Is the Earth warming? And are we responsible? Among the American public, a thin majority are convinced that the planet is changing. Even more believe that we humans have nothing to do with this change.

Over the next fifty to a hundred years, global change has to be the greatest economic and environmental threat facing the planet. For example, if glaciologists are even partly right, many of the world's coastal cities will be in trouble because of sea level rise caused by melting ice sheets and the warming ocean, and millions of people will be environmental refugees, displaced from the deltas of the world's major rivers. Yet we continue to be bogged down by the dilatory tactics of those who deny that the threat even exists. Of the two major political parties in America, one holds that global change is nothing to be worried about and may even be a fraud, and the other holds that it's probably real, but not real enough for immediate, concrete action.

These are strange times. Never in our nation's history has a question of science become so politically polarized as to become a partisan issue. Never in our history have the basic tenets and procedures of science been so widely questioned by the news media. And all this because the scientific community is the bearer of the bad news that we as a society are producing too much carbon dioxide—a conclusion that is viewed as a threat to the future of the energy industry.

For the world's scientific community to become galvanized on a socially important and politically charged issue such as climate change is an extraordinary event in itself. Traditionally scientists have kept to themselves, kept their noses to the grindstone and their eyes to the microscope, studying things that industry needed, or worrying about (according to Charles Wilson, former president of General Motors) why the grass is green and the sky is blue! This foray into the politicized world of climate change debate by a large portion of the world scientific community has not always gone well. The United Nations Intergovernmental Panel on Climate Change, the group responsible for the major reports documenting the threat, has become the lightning rod for a small army of deniers. Scientists, usually happy to publish articles in academic journals largely unnoticed by the public, now find themselves the targets of politically based criticism.

While some industrialized nations such as the United States have been slow or reluctant to recognize the threat of climate change, other, smaller countries already threatened by global warming have taken action. Bhutan, a tiny country of 700,000 souls between India and China, may be the most global-change-conscious of all. Bhutan is taking major planning steps to move villages as their glacial water supply dries up, while next door, China is still in a state of denial concerning imminent groundwater and inundation problems in Shanghai from the sea level rise.

A note about terminology: The phenomenon of global changes related to increased atmospheric carbon dioxide and other greenhouse gases has been widely referred to as *global warming*. However, within the scientific community there is concern that referring just to the warming downplays the large number of other events that are caused by the warming. Some scientists prefer the term *global climate change*. This term suffers from a similar shortcoming, in that sea level rise, retreating glaciers and ice sheets, melting permafrost, plant and animal displacements and extinctions, and warming oceans are not climate changes. We use the term "global climate change" in our title because we believe it will be widely recognizable by our readers. However, in the book we will usually refer to the overall phenomenon as *global change* and use the term *global warming* mainly when we are discussing temperature changes per se.

We wrote this book because we felt that the general public had few resources that explain briefly, simply, and in layperson's language the science of global change. It is intended as a primer of global change. It is not illustrated by the usual graphs, maps, and tables, but instead by the batiks on silk of Mary Edna Fraser. We hope that her batiks will make the book palatable to the most scientifically challenged among us, and bring it to the attention of an audience that doesn't always see books on science.

Throughout the book, we take pains to point out that there are many uncertainties about future climate change and that some of the supporting data are limited. Although the basic assumptions are solid, there is, as in the steering wheel of an old car, some play in them. We even argue that mathematical models, the main basis of many of the important conclusions, must be viewed with caution (because they are based on the same assumptions with play!). At the end of each chapter we have a section where we state some of the deniers' beliefs and proceed to explain why they (usually but not always) are wrong.

Orrin Pilkey, one of the authors, is a professor emeritus from Duke University whose long career in marine and coastal science has led him to produce many publications in his realm of knowledge. Keith Pilkey, his son and co-author, is an attorney with the Social Security Administration who has long been interested in the role of skeptics in our society's global change debate. Mary Edna Fraser is an artist from Charleston, South Carolina, specializing in coastal scenes. Her career has led her to photograph the eastern seaboard of the United States and coastlines abroad, often from the cockpit of planes she is piloting with an instructor by her side. She translates the photographic images onto silk in the ancient art of batik.

ORRIN AND KEITH PILKEY would like to thank a number of people who helped them to understand some aspects of global change far from their own specialties and to navigate the shoals of climate change denial. We are particularly grateful to Paul Baker, Bruce Corliss, Fred Dobson, Philip Froelich, Peter Haff, Miles Hayes, Duncan Heron, Joe Kelley, Susan Lozier, Stephen Mako, Brad Murray, Bill Neal, and Stan Riggs; Dana Beach of the Coastal Conservation League; and Todd Miller of the North Carolina Coastal Federation.

We are particularly grateful to Reynolds Smith, who came up with the idea for this book. We thank Sharlene Pilkey for spending many hours doing internet research on a variety of topics. We thank Norma Longo for helping us at great length with organization and references, proofreading, and copyediting. We also received copyediting help from Reynolds Smith, Cecelia Dailey, Mary Edna Fraser, and Gina Longo. We are very grateful for the backing of all of our families.

MARY EDNA FRASER would like to recognize the help of the National Aeronautics and Space Administration (NASA), the National Oceanic and Atmospheric Administration (NOAA), the Heinz Center, the North Carolina Museum of Natural Sciences, Emory University, the McKissick Museum, the Flaten Art Museum, the McClellanville Arts Center, the Hickory Museum of Art, the University of Oklahoma, the Barrier Island Center, the Florence Museum of Art, Science and History, and the Peabody Essex Museum. Also, she is grateful to her patrons, including Mark and Clarinda Abdelnour, Ingrid Abendroth, Norman and Chris Lorusso, Rick McKee and Tara Lowry, John and Betty Jean Payne, Norton and Mindy Seltzer, Mary Lou Stevenson, and Marty and Cathy Wice. She thanks her assistant, Cecelia Dailey, her intern Chase Cribb, Reynolds Smith, Norma Longo, Jill Ewald and Mengchi Ho, who helped with edits; Rick Rhodes and Tim Steele, for photography and digital files; her husband John Sperry, who offered untiring support; and the members of her family, who made her life as an artist possible. Mary Edna appreciates the decades of friendship and education from her "boss" Orrin Pilkey.

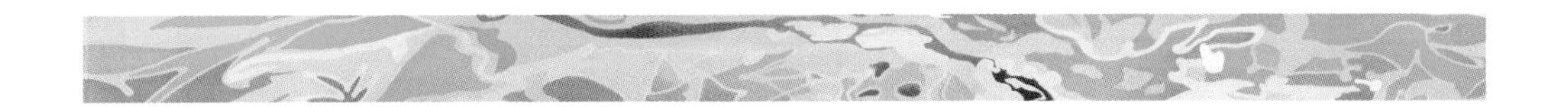

GLOBAL CHANGE AND THE GREENHOUSE EARTH ✳ 1

The Greenhouse Effect through the Ages

Ever since water accumulated to form the ocean, not long after the Earth formed 4.6 billion years ago, the level of the sea has been moving up and down. It was only in recent millennia, however, that such changes have affected human beings. For example, a few miles off the coast of Maine, fishers have been trawling up spear points, arrowheads, and other stone implements from a submerged village site, which was occupied eight to eleven thousand years ago. The people who lived at this site, now under a hundred feet of water, had to pick up and move inland as the sea level rose and the shoreline moved past their village. Almost certainly it wasn't just the gradual flooding by the rising sea that forced them to flee. Probably it was a storm or two that penetrated further inland than usual, or perhaps their drinking water became too salty because of the higher sea level. The effort required to move a prehistoric Native American village inland is a far cry from what would be required to relocate today's New England coastal settlements.

It seems that nothing is new under the sun. The temperature of the Earth's atmosphere, just like the level of the sea, has varied considerably over time. Most of the changes have occurred because of variations in the amount of heat received from the Sun, which are related to solar activity and the orientation of the Earth relative to the Sun.

Atmospheric temperature is also affected by the concentration of what scientists have termed greenhouse gases in the atmosphere. The greenhouse effect makes life on Earth possible. As solar radiation warms the Earth's surface, a portion of the Earth's atmosphere acts like a greenhouse and retains heat that would otherwise be lost back to space.

At the time of the dinosaurs, the Earth's temperature was much warmer. Now, 65 million years after the last dinosaur died, we have reached a very cool period known as the Ice Age. Even though we are currently in an interglacial time, between advances of the great ice sheets, it is a time that is too cool for cold-blooded dinosaurs in most of the places where they previously lived.

Sea level and atmospheric temperature are often related. When global temperatures cool, continental scale glaciers can form, and they can contain so much of what was once the ocean's water that they cause drops in sea level as large as four hundred feet. There have been at least seven major fluctuations of the sea over the last two million years.

At present the Earth's huge and ever growing human population is affecting the climate through (1) the emission of gases that cause the Earth to act like a greenhouse; (2) the discharge of high concentrations of atmospheric particles called aerosols; and (3) changes in land use. All three are interrelated and work in tandem to change the Earth's temperature.

A real greenhouse works differently from the Earth's atmosphere. Glass-enclosed greenhouses prevent winds from dispersing the heat created by the Sun's rays. Thus the interior space retains heat. The atmospheric greenhouse works because the greenhouse gases allow short-wavelength sunlight to pass through the atmosphere. This radiation is converted to heat, which warms the Earth's atmosphere, causing it to emit longer-wavelength infrared radiation back into space. The greenhouse gases absorb some of this infrared radiation heading out into space, which in turn heats up the Earth's atmosphere even more. Eventually the infrared rays escape the Earth, but the more greenhouse gases there are in the atmosphere, the more radiation is trapped, resulting in increased temperatures.

While the greenhouse effect is natural and necessary for human life, the fundamental problems are that human activities have created an excess of greenhouse gases in the atmosphere, and that the enhanced greenhouse effect has brought about global warming at a rapid and accelerating rate.

The Greenhouse Gases

So, what exactly are the major greenhouse gases? The principal greenhouse gases whose presence in the atmosphere causes warming are water vapor (H_2O), carbon dioxide (CO_2), methane (CH_4), and nitrous oxide (N_2O). All the greenhouse gases are "trace gases": that is, they all make up a very small part of the atmosphere. Carbon dioxide is usually measured in parts per million (ppm). Present measurements indicate that there are around 390 molecules per million molecules of air. Methane and nitrous oxide are measured as parts per billion (ppb). Even water vapor is a minor constituent of the atmosphere—at most a few percent.

Yet water vapor is responsible for most of the greenhouse effect, accounting for between 36% and 66% of the warming (average of 60% globally), but it is more or less a long-term constant in the greenhouse equation. That is, it is not directly responsible for the global climate change that the Earth is experiencing. Unlike the other greenhouse gases, which are uniformly distributed in the atmosphere, water vapor

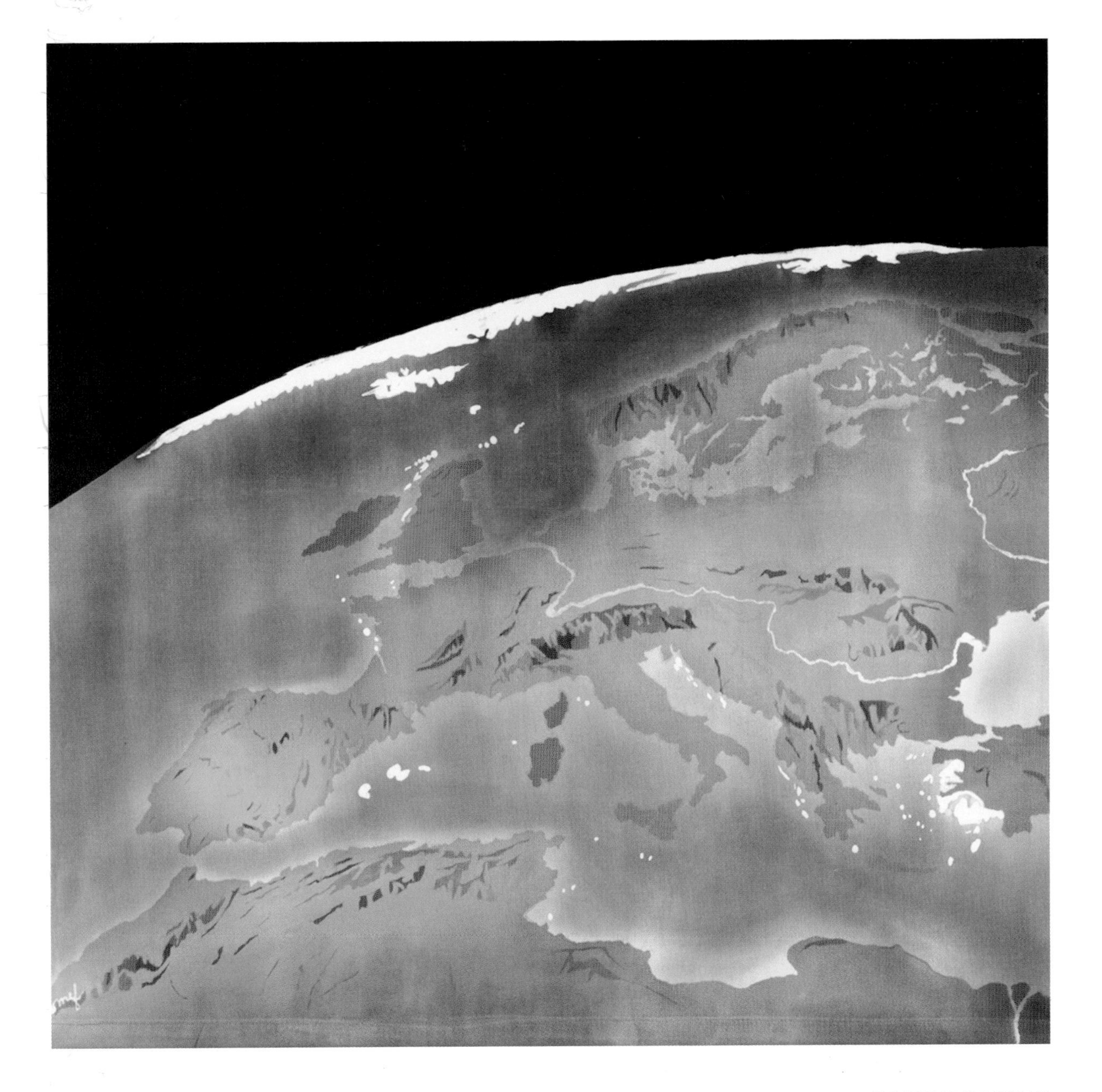

EARTHSCAPING

The Earth is warming: of this there is no doubt. The evidence that is there for all to see includes the melting ice sheets in Greenland and Antarctica, the melting mountain glaciers, the shrinking of both permafrost and Arctic sea ice, the rising sea levels and the warming oceans. Global climate change is a problem that our society must address sooner rather than later.

concentrations vary widely both in space and time.

A warm atmosphere is able to hold more water vapor. Warming of the atmosphere causes an increase in water vapor content, which in turn leads to more warming. Increased warming leading to increased warming is what is called a positive feedback. However, increased water vapor can result in decreased warming (a negative feedback) if more cloud cover is formed, because clouds reflect solar radiation, a distinct cooling effect. Clearly the role of water as a greenhouse gas is a complicated one and presents a problem for predictive mathematical models (as discussed in chapter 3).

The global warming potential of a gas (GWP in the accompanying table) is a measure of how much a gas is estimated to contribute to the greenhouse effect. The global warming potential depends on both the efficiency of the molecule as a greenhouse gas and the length of time it remains in the atmosphere. Both factors are summarized in the table, in which CO_2 is given an arbitrary value of 1 for the purpose of comparing it with other gases over a period of twenty years. The right-hand column in the table indicates that methane is 72 times more powerful as a greenhouse gas than CO_2 and nitrous oxide 289 times more powerful. Because CO_2 has a higher concentration than most other gases, its impact on global warming is largest, even though the other gases have a larger global warming potential. That is, a molecule of methane is a much more powerful greenhouse gas than CO_2, but it has less of an impact because a methane molecule resides in the atmosphere for only twelve years (this is its "lifespan," as shown in the middle column of the table) compared to thousands of years for a CO_2 molecule.

Carbon dioxide is the principal greenhouse gas villain because it is the gas produced most abundantly by human civilization in the modern era. Human activities produce eight billion tons of CO_2 per year compared to the largest natural source, volcanic activity, which accounts for less than a third of a billion tons. During the cold times at the height of the last ice age, the CO_2 content of the atmosphere was 180 ppm. The concentration has since progressed from 280 ppm in the period preceding the Industrial Revolution (the eighteenth and

Table 1. Global Warming Potential (GWP) over Twenty Years

Greenhouse Gas	*Concentration (%)*	*Lifespan (Years)*	*GWP*
Carbon dioxide (CO_2)	77	1000s	1
Methane (CH_4)	14	12	72
Nitrous oxide (N_2O)	8	114	289
Chlorofluorocarbons (CFCs), etc.	1	1000s	1000s

nineteenth centuries) to a present-day 390 parts per million—higher than it has been for 650,000 years (based on the study of air bubbles in ice core layers from Greenland). Based on measurements taken at the top of Mauna Loa in Hawaii, the rate of increase of CO_2 is accelerating and now stands at about 2 parts per million per year.

A decade ago a common assumption was that to hold back major and irreversible climate changes, excess production of CO_2 should be kept below 550 parts per million in the atmosphere (nearly two times preindustrial concentrations). The bill in the U.S. Congress known as the Waxman-Markey bill aims for 450 parts per million, and that target could require an 80% reduction in emissions by mid-century. Currently the atmospheric CO_2 concentration is approximately 390 parts per million. The NASA climatologist James Hansen argues that 350 parts per million is the concentration we should be aiming for and that anything higher than that (including the present-day concentration) takes us beyond a tipping point where irreversible changes will occur (for example, runaway melting of the ice sheets and rapidly rising sea levels). Hansen's call for a reduction in CO_2 to 350 parts per million in order to "preserve a planet similar to that on which civilization developed and to which life on Earth is adapted" has led to the creation of 350.org, an environmental organization headed by the author Bill McKibben.

That there is disagreement as to the precise parts-per-million number beyond which irreversible climate change occurs does not in any way detract from an important point: reduction of CO_2 in the atmosphere is an absolutely essential goal.

How do we know that the increase in CO_2 is not simply part of a natural cycle, as is commonly argued by climate change deniers? The best evidence that the CO_2 increase results from the burning of fossil fuels is the carbon isotope mixture in the atmosphere. Isotopes are two different forms of the same element, and carbon has two stable isotopes, carbon 13 and carbon 12. Plants prefer to take up a lighter mix of isotopes than is present in the atmosphere, that is, an isotope mix richer in carbon 12. Most coal and oil is derived from plants, so as these fuels are burned they contribute back to the atmosphere a relatively light mixture of carbon atoms. That the atmosphere is getting lighter in terms of its carbon isotope mix (with more carbon 12) is a measure of the contribution of fossil fuel burning.

Methane, the second-most significant gas for global warming after CO_2, has a total greenhouse effect about one-third that of CO_2. As can be seen from table 1, methane is a more powerful greenhouse gas (global warming potential of 72, compared to 1 for CO_2) but a less abundant one (14% of total greenhouse gases), and molecules remain in the air for only a short time (average lifespan of twelve years). Approximately 55% of the annual methane emissions into the atmosphere are from anthropogenic sources, the most important of which are energy

SLOPES OF MAUNA LOA, HAWAII

Charles David Keeling, a visionary scientist from Scripps Institution of Oceanography, measured the carbon dioxide (CO_2) in the relatively unpolluted atmosphere at the top of Mauna Loa beginning in 1958. Mountain-top measurements now show that CO_2 concentration in the atmosphere has steadily increased and is nearly 390 parts per million compared to 315 parts per million in 1958.

production and the raising of ruminants (livestock). Natural methane emissions are primarily from wetlands, but agricultural sources other than ruminant livestock, e.g., rice paddies, also contribute to methane gas emissions.

During the last glaciation, methane was found at concentrations of 400 parts per billion (ppb). After most of the ice left, it rose to 700 ppb, and now, after the Industrial Revolution, methane has reached concentrations of 1,500 ppb.

The cycle of methane emissions is not as well understood as that of carbon dioxide, but climate change itself may soon unleash vast natural reserves of methane and thereby dramatically amplify the greenhouse effect. One particular area of concern is the thawing permafrost of the Arctic, especially in the vast tundras of Siberia. Melting of the ice contained in the soils will enhance the bacterial degradation of plant matter long stored in the soils, releasing not only potentially large volumes of methane, which is a byproduct of bacterial decay, but also long-dormant carbon dioxide molecules. The 2008 yearbook of the United Nations Environment Program warned that "methane release due to thawing permafrost in the Arctic is a global warming wild card," meaning that the potential volume of CH_4 release is very large and very damaging, but when it will be released remains a mystery.

The second, even larger potential source for methane gases is the methane hydrates or methane ice stored in the deep sea. On a local scale the massive oil spill in the Gulf of Mexico in 2010 released a lot of methane ice. In the cold Arctic, methane ice may be found in ocean waters as shallow as one hundred meters. Methane hydrates are huge deposits of methane produced by degrading organic matter that are maintained in a frozen, immobile state by the cold temperatures and high pressures at the bottom of the deep-water column (usually greater than one thousand feet). These methane ice deposits, which hold the potential for catastrophic global changes, are found on many continental margins, including those of eastern North America. They are also an important exploration frontier for oil companies, which see a huge energy potential in the deepwater deposits. Smaller but significant volumes of methane hydrates exist on land, beneath permafrost in the Arctic.

As the ocean waters warm or as warm currents change their trajectory (often a result of climate change), more and more hydrates will be released from the shallower deposits. Plumes of methane gas bubbles have been observed coming from the sea floor on the West Spitsbergen Arctic continental shelf, and very high concentrations of methane in seawater have been observed recently on the East Siberian continental shelf. According to Natalia Shakhova of the University of Alaska, eight million tons of methane escape every year off Siberia, an amount equal to what was previously assumed to be the total amount of methane released from all the oceans. In this case the methane is probably being released

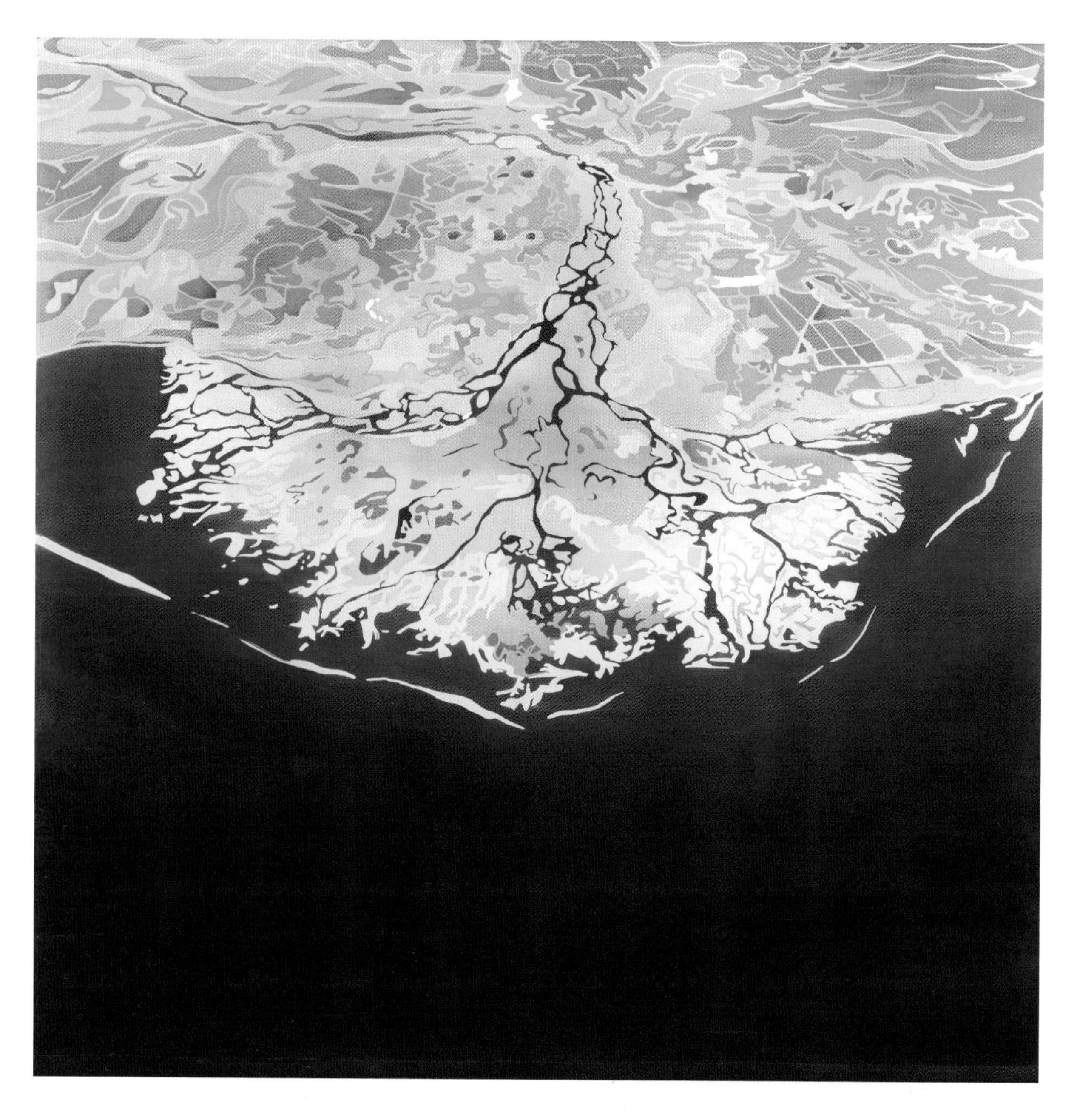

SELENGA DELTA, RUSSIA

The beautiful Selenga River Delta in Siberia extends into Lake Baikal, the largest freshwater lake by volume in the world. Melting of the permafrost on the delta and in this region will add significantly to the methane concentration in the atmosphere.

because the Arctic current has warmed over the past three decades, causing the release of methane by breaking down methane hydrate in the sediment beneath the seabed. This Siberian discovery is an example of the evolving nature of our knowledge of greenhouse gas emissions. We still have much to learn.

The sudden release of massive amounts of methane from marine methane ice is the suspected cause of two of the Earth's major extinction events. The Paleocene-Eocene Thermal Maximum of 55 million years ago led to the extinction of numerous marine and land-based organisms. In this instance the collapse of methane ice deposits seems highly probable as the cause of the occurrence of spectacular atmospheric warming, which took perhaps 100,000 years to recover from. The much larger Permian-Triassic Extinction Event 250 million years ago resulted in the extinction of 70% of all land vertebrate species. Although the cause is much less certain because the event happened so long ago, conceivably methane ice melting and a runaway greenhouse effect were behind it as well.

You can find dramatic videos on YouTube of scientists from the University of Alaska, Fairbanks, punching holes in frozen lakes and lighting up the methane released by melting permafrost, gas which is temporarily trapped by the ice.

Nitrous oxide is currently responsible for about 6% of the heating caused by greenhouse gases. It is a pollutant from industry and particularly from agricultural fertilizers. It is also released naturally from soils and from the oceans. Besides its role as a greenhouse gas it reacts with and destroys ozone in the atmosphere.

Freons are a number of compounds that have no natural source but are now in the atmosphere. These include chlorofluorocarbons (CFCs) used in refrigerators; their replacements, hydrofluorocarbons (HFCs); nitrogen trifluoride from flat-screen televisions; and halons from fire extinguishers. All are very powerful greenhouse gases (high global warming potential) but they occur in such small quantities in the atmosphere that to date their impact on warming has not been important.

The following list is a generalized summary of the global sources of greenhouse emissions. Almost all the listed sources produce emissions primarily through the burning of fossil fuels such as coal, oil, and natural gas. Only one, land use change, produces emissions when the carbon-sequestering abilities of the land are altered.

Agriculture 13%
Industry 19%
Transport 13%
Residential and commercial buildings 8%
Energy production 26%
Land use change 18%
Waste management 3%

There are other ways to analyze sources of global emissions. For example, according to the United Nations Food and Agriculture Organization, the meat in our diet is a major source of greenhouse gases.

The preparation of chicken, pork, and beef produces more greenhouse gases than the transportation or industrial or residential sector. Of the 36 billion tons of CO_2 and other greenhouse gases released annually, 18%, or 6.5 billion tons, is from meat production. Beef production is the largest villain and it works like this: 40% of the emissions are from reduced CO_2 absorption caused by the loss of the plants that cattle eat; 32% is from the gases that cattle emit and the emissions from their waste; 14% is from fertilizer production; and 14% is from general farm activities. According to an article in *Scientific American* by Nathan Fiala, the emissions attributable to the meat in a hamburger from Burger King are equivalent to the emissions from driving a small car for ten miles. Consider that in addition, modern meat production typically involves transporting the product far from the slaughterhouse, and that Whopper you consumed for lunch is increasing more than just your waistline.

The Role of Aerosols

Aerosols are tiny particles or droplets suspended in the atmosphere. Much remains to be learned about them, but it seems that most atmospheric aerosols are purely natural, derived from natural processes such as oceanic salt spray, forest fires, and dust storms. Volcanic eruptions, such as the eruption in 1991 of Pinatubo on the island of Luzon in the Philippines, can temporarily inject huge amounts of aerosols into the atmosphere. Human beings contribute black carbon aerosols through slash-and-burn deforestation, and they produce soot, sulfates, and nitrates in the form of smog from air pollution and dust created by agriculture, desertification, and other activities and events on land. Some aerosols have a regional cooling effect, as they reflect solar radiation back into space. And cooling from a major volcanic eruption will usually last a year or two. Industrial effluents also include such aerosols, some of which cause cooling (sulfates) while others cause warming (soot).

Some aerosols are also responsible for cloud formation, and the amount of aerosols in the atmosphere determines the type of cloud. High concentrations result in bright, white clouds, which are very effective in reflecting sunlight (thereby cooling the Earth). Darker clouds, which form when aerosols are less concentrated, will allow more of the sun's energy to reach and warm the Earth.

Satellite observations have shown that the paths of ships in the open ocean can sometimes be seen as spectacular, long, thin lines of clouds, produced as the sulfur dioxide from smokestacks is emitted in the form of sulfate aerosol particles which lead to cloud formation. This is a clear indication of the importance of fossil fuel burning in forming aerosols.

The overall impact of aerosols is to cool the Earth, more in the northern hemisphere than in the southern hemisphere. The Intergovernmental Panel on Climate Change (IPCC) estimates that perhaps aerosols can reduce the warming effect of CO_2 by as much as one-third (which is why injection of aerosols into the atmosphere has been suggested as a means of artificially cooling the Earth, as discussed in chapter 9). Industrial polluters have in effect been injecting aerosols into the atmosphere for years, and thus there may already be a cooling effect. One characteristic of aerosols is that they have a shorter residence time in the atmosphere than greenhouse gases do.

A recent study led by Nadine Unger of NASA's Goddard Institute for Space Studies revealed the complex interaction of greenhouse gases and aerosols. Rather than focus solely on the volume of gas emissions in global warming, the study took into account the combined atmospheric heating effect of the gas and the cooling effect of the aerosols produced at the same time by various industrial sectors. They concluded that on-road transportation (cars, buses, trucks) is the greatest net contributor to global warming, followed by burning biomass for cooking foods and raising animals for food (particularly methane-producing cattle).

The Role of Land Use Changes

Approximately one-third of CO_2 emissions over the past two centuries has come about because of global land use changes. The most important of these is deforestation. Removing natural forests reduces the amount of CO_2 that will be absorbed by the plants, thus increasing the overall CO_2 concentration in the atmosphere. It also exposes soil to decomposition, further adding to the emissions.

In the past century the causes of deforestation have shifted. Subsistence activities and government development projects have given way to mining and large-scale ranching and farming. There is no agreement on how much rainforest has disappeared, and of course some of it has been and is being replanted after deforestation. The most extreme deforestation is probably in Haiti, where 1% of their forest remains. Just 10% of West Africa's coastal rainforests and perhaps 12% of the rainforests of South Asia remain. Brazil has declared deforestation to be a national emergency.

In the typical process of deforestation, greenhouse gases are released by the burning of the felled forest. This burning, of course, also releases aerosols, which will have a cooling effect, as discussed above. Finally, land use change, particularly deforestation and agriculture, decreases the albedo of the Earth's surface (the ratio of the light reflected by a planet to that absorbed by it). This causes more heat to be taken up by the Earth.

AMAZON RIVER

Deforestation in the Amazon forest increases the CO_2 concentration of the atmosphere, because removal of trees reduces the amount of CO_2 to be absorbed by plants. On a global scale deforestation is a major source of the recent increase in atmospheric CO_2.

The Greenhouse Turmoil

There is a growing sense of public concern and skepticism about the science of global change. A good bit of disagreement among scientists about the details of greenhouse warming is sometimes mistaken for a lack of consensus. This misunderstanding stems from a larger one about the adversarial nature of science. A paper in *Science* magazine in 2010 signed by fifty-five members of the National Academy of Sciences notes that "scientists build their reputation . . . not only for supporting conventional wisdom but even more so by demonstrating that the scientific consensus is wrong." Criticism and discussions between scientists are the way science is conducted and not an indication of weakness or uncertainty. Honest criticism and debate are the hallmark of good science. Criticism strengthens science.

The academy members also note that "many recent assaults on climate science . . . are typically driven by special interests or dogma, not by an honest effort to provide an alternative theory that credibly satisfies the evidence."

Myths, Misinterpretations, and Misunderstandings of the Deniers

The following are examples of commonly made misstatements about global change, perpetuated by global warming deniers. We will list these myths and rebut them (and occasionally agree with them) at the end of most chapters throughout this book.

MYTH: *If we can't predict weather accurately for the next five days, how can we predict the climate decades and even a hundred years from now?* During a debate on global warming in 2010 with Robert Kennedy Jr., the CEO of Massey Energy, Don Blankenship, shared this insight regarding global warming: "It's a hoax because clearly anyone that says that they know what the temperature of the Earth is going to be in 2020 or 2030 needs to be put in an asylum because they don't." Weather is different from climate. Weather is difficult to predict, in part because it is a chaotic system and involves looking at very short-term events. We can say with some accuracy that fifty years from now the winter climate in Chicago will be colder than the summer climate, but we cannot be assured that our Chicago weather prediction six days from today will be accurate. Climate is a long-term average of weather, which smooths out regional or short-lived weather extremes. Nonetheless, concern about the accuracy of long-term projections is real, and these projections must always be viewed with caution.

MYTH: *In the geologic past, global warming was followed by an increase in CO_2 levels (not the other way round), so CO_2 increases do not cause warming but instead are caused by warming.* Representative Joe Barton (noted for receiving funding from the political action committee of the fossil fuel giant Koch

Industries, but not necessarily for his understanding of climate science) posted this statement to his website: "An article in *Science* illustrated that a rise in carbon dioxide did not precede a rise in temperatures, but actually lagged behind temperature rises by two hundred to a thousand years. A rise in carbon dioxide levels could not have caused a rise in temperature if it followed the temperature." In the past, when temperature changes were due to variations in solar radiation and other factors (and not due to human activity), the statement may have been accurate. Past warming of the oceans may have been initially started by something other than CO_2, and since warmer waters hold less gas, CO_2 was released and amplified the warming through the greenhouse effect, thus adding to global warming. In the last few decades of the twentieth century, however, CO_2 and temperature increases occurred together.

MYTH: *Water is far more abundant and a far more important greenhouse gas and therefore accounts for most of the greenhouse effect.* Water in the atmosphere, as discussed above, varies widely from place to place and from time to time but can be considered a constant as far as global warming is concerned. It is not one of the major factors driving global warming. Seasonal or weather changes in water concentration in the atmosphere are short-lived. If there's too much water, rain will reduce it. If there is not enough water, evaporation from the ocean will bring it back up.

MYTH: *CO_2 makes up 0.0387% of the volume of the atmosphere and therefore must be insignificant.* Even at its small level of concentration in the atmosphere, theoretical considerations, lab studies, and field observations all show that CO_2 is an important cause of global warming.

THE IMPACT OF GLOBAL CHANGE ✳ 2

Warming

The evidence that the Earth is warming consists of its rising sea levels, warming atmosphere and oceans, and widespread melting of sea ice, ice sheets, glaciers, and permafrost. Short-term temperature numbers over the last century were measured by thermometers, of course, but longer-term records are mostly derived by interpreting ice cores from ice sheets and glaciers, deep-sea sediment cores, tree rings, and corals.

Atmospheric temperature measurements since the beginning of the twentieth century indicate that the Earth's average temperature has increased from 13.5 to 14.5 degrees Celsius (56 to 58 degrees Fahrenheit). Temperature warming varies with the season and with latitude, as illustrated by the warming trends in Alaska. Between 1949 and 2008, summer temperatures in Alaska have risen by an average of 1.2 degrees C (2.1 degrees F), fall temperatures by 0.5 degrees C (0.9 degree F), winter temperatures by 3.3 degrees C (6 degrees F), and spring temperatures by 1.9 degrees C (3.5 degrees F). Overall the average temperature has risen by 1.7 degrees C (3.1 degrees F), a number that is typical of the rapidly warming high latitudes. The temperate and tropical zones are warming more slowly.

Although the greenhouse effect is global, each of the continents has a slightly different average temperature curve for the twentieth century. This is because a myriad of processes other than greenhouse gases and solar radiation have an impact on temperatures. These include ocean currents, wind patterns, the presence of ice, the distribution of land areas, vegetation patterns, and volcanic eruptions. The most marked regional changes have occurred in the Earth's polar regions. Overall the Arctic has warmed in recent decades, even though temperatures dropped dramatically in the early 1960s after having risen during the 1930s and 1940s. The Arctic has warmed more than the Antarctic because there is more land area in the northern polar regions to absorb solar radiation. Land areas warm twice as fast as the upper layers of the sea, in part because the ocean loses much more heat by evaporation.

Local effects are quite variable, however, and local temperature curves may vary significantly from the global mean curve. Some areas have actually experienced temperature drops and others have remained the same, but most have experienced a warming atmosphere.

Until recently Antarctica appeared to be the only continent that was not warming. Because of the size of the continent and the sparse number of weather stations, this was never a certainty. In 2009 two separate, peer-reviewed reports indicated that the Antarctic continent had warmed considerably over the preceding fifty years. One study based on ice cores on the Antarctic Peninsula showed a warming of 2.7 degrees Celsius (4.9 degrees Fahrenheit) during that period. A second study based on satellite observations on the West Antarctic Ice Sheet suggested a warming rate of 0.1 degree Celsius (0.2 degree Fahrenheit) per year over the preceding decade. Both these rates of temperature increase are much higher than the global warming average.

The atmosphere is not the only Earth body that is warming. In fact, 80% of the Earth's heat gain over the last fifty years has been stored in the uppermost two thousand feet of the ocean and not in the atmosphere, which accounts for the other 20%. The difference is due to the far greater capacity of ocean waters to absorb heat. The physics of air, water, and heat dictate that the increase in heat content of the oceans produces a much smaller rise in temperature than a similar amount of heat absorption in the atmosphere. Nonetheless the volume of water in the oceans is so immense that a small rise in temperature can cause a significant expansion of the water, which has been the cause of most of the sea level rise that we have experienced over the last hundred years. This is discussed further in chapter 6.

The bottom line:
the entire globe is warming.

Global Climate Changes

The impact of global warming on local and even regional scales is difficult to predict and remains in the realm of educated guesswork. It is certain that the Earth's ecosystems will shift as new species are added (or subtracted), but uncertainties abound. Unknowns that could produce surprises include changes in major ocean currents (such as the Gulf Stream), patterns of cloud distribution, and storm patterns. How the human population grows and shifts about and how land use patterns change will also greatly influence future climates.

In a very general way, wet regions will become wetter and dry regions will become drier. As the atmosphere warms, it will contain more water vapor, leading to more rainfall globally. Also, as the air gets warmer, evaporation will increase, snatching water up from the land. Probable regional trends include increasing aridity around the Mediterranean, South Africa, Southern and Central Australia, Chile and Argentina, Pakistan, Mexico, and the south-western United States. Widespread desertification has already occurred in North Africa, in part because of overgrazing by sheep and goats. Similar poor land use practices, along with climate change in Spain, threaten to turn one-third of the country into a desert. At the other end of the spectrum, Canada, the eastern United States, much of Russia, the entire Arctic region, and northern Europe are expected to be wetter.

Both good and bad things will happen in North America. Warmer temperatures will expand agriculture to more northern latitudes. Increased atmospheric CO_2, plus a longer growing season, may aid crop and forest growth. Food prices may therefore drop and food exports may increase. Lower heating bills, longer construction seasons, and longer seasons for warm weather recreation will occur. In high latitudes new ports may open, and the Northwest Passage may become a reality.

The negative side of things begins with sea level rise that will certainly threaten North America's major coastal cities, especially along the low-lying East and Gulf Coast coastal plain. The cost of responding to sea level rise in New York, Boston, Philadelphia, Washington, Miami, and New Orleans will be monumental and will likely be the nation's highest-priority global change response in the not-too-distant future. In the Southeast, barrier islands may be destroyed or become uninhabitable. In North America arid lands will expand in the Southwest region. Tundra in the subarctic will be reduced, perhaps by as much as by two-thirds by 2100, releasing abundant methane into the atmosphere and creating serious problems for subsistence societies such as the Inuit and Yupik of Canada and Alaska, and the indigenous peoples of Siberia. Decreased winter snow packs will significantly reduce the water supply of western North America, leading to political conflict over water allocation. Forest fires will increase in number and size in the West. The need for air conditioning will increase, and the time span for winter sports will be reduced.

Forest Fires

An increase in the size and frequency of forest fires is predicted for a number of regions that already experience them. This is a response to a warming atmosphere, drying forest soils, a long history of fire suppression (allowing undergrowth to expand), longer fire seasons (the fire season in the western United States has already increased in length by more than two months), and changes in the size and timing of melting winter snow accumulations. Most of the world's large forest fires occur in Siberia and northern Canada, but these don't receive the public attention of fires in more extensively developed areas. In the summer of 2010 Russian forest fires extending from Siberia to the outskirts of Moscow did attract international attention. The fires were more extensive than ever, causing the loss of a number of towns and villages. Forests in the western United States, where the area burned each year has increased dramatically in the past two decades, are particularly susceptible to fire. Recent examples of large fires in developed areas, all in 2009, include fires in Greece (which burned 21,250 hectares, or 52,500 acres); the Station Fire in Los Angeles County (56,650 hectares, or 140,000 acres); and the Black Saturday bushfires in southern Australia (445,000 hectares, or 1,100,000 acres), which killed 173 people. All these fires damaged or destroyed many buildings.

Other forested regions that will be susceptible to large fires in the future include parts of South Africa, much of Europe, the Mediterranean rim, and the Amazon Basin.

These large fires create serious problems for wildlife and are major sources of pollution, as post-fire rains cause extensive erosion on ash-covered slopes, polluting nearby streams and rivers. All fires in nature, plus those created by human beings in the deforestation process, provide a CO_2 contribution equal to 50% of the CO_2 contribution from fossil fuels. In the Pacific Northwest and other parts of the world as well, brush and dead trees are often cleared from forests to reduce the potential for future fires. Unfortunately this reduces the carbon sequestering effect of forests and thus effectively increases the CO_2 content of the atmosphere.

Desertification

Western civilization, its wealth and power symbolized by the busy bulldozer, can't seem to resist the temptation to confront nature in places where we arguably don't belong. Deserts are such places: only a small number of people can eke out a living in deserts unless they are supported by massive technology such as dams and irrigation systems. In his book *Cadillac Desert*, Marc Reisner recounts the settling of the deserts in the western United States, a process leading to the current situation, in which large amounts of money are infused into these regions, resulting in dwindling water supplies and increasing conflicts between water users.

Desertification—the deterioration of land and the reduction of vegetation and soil in arid areas—results from climatic variations as well as human activities such as those described by Reisner. Desertification is expected to spread in the American West and likewise in Australia. The Mozambique coastal zone in East Africa is at particularly high risk for desertification, in part because local inhabitants have destroyed mangrove forests along the shoreline. Desertification is also a strong possibility in a broad sub-Saharan band across the African continent. Over a number of centuries 28% of China has been affected by desertification, much of it caused by deforestation and overgrazing. Changing weather patterns, which contribute to

WILSONS PROMONTORY, AUSTRALIA

Forest fires such as these shown on Wilsons Promontory along the southwest coast of Australia are expected to become larger and more frequent as global warming increases. The Russian and Bolivian forest fires in 2010, the Australian bush fires in 2009, and the Alaskan tundra fires of 2007 are examples of recent fires that were the largest ever recorded.

desertification, will inevitably result in increased human conflict as populations struggle for control of dwindling natural resources. In an article in the *Atlantic* in 2007, Stephen Faris linked the conflict in Darfur to desertification caused by climate change. If this was one of the first climate change–related wars, it will certainly not be the last.

Up to the present day, human activities such as overgrazing and poor farming practices have been largely responsible for desertification in already arid regions. The same is likely to be true in a time of global change. Humans will play the major role in causing desertification, but higher temperatures and perhaps other climate changes will ease the way.

The saga of the Murray-Darling River Basin of southeastern Australia, named after its two main rivers, is likely a peek into the future of a warming world in an arid region. That much of the basin, which is about the size of France and Spain combined, was not prime agricultural land was recognized early on. In 1865 a surveyor on horseback, George Goyder, mapped a line in the basin, now called the Goyder line. It marked the boundary between arable grassland and sparsely vegetated bush country not suitable for farming.

About the time of the First World War, settlers with government support began to cross the Goyder line. Native vegetation was essentially removed, crops that required large amounts of water were grown, and water from the rivers was over-allocated. Soils became salinized as irrigation water dissolved salts within and below the soil layer; the salts then made their way to the soil surface, leaving telltale white soil patches. The most water-hungry crops were cotton and rice. Livestock and dairy cattle also consumed much water; a thousand gallons of water are required to produce one gallon of milk. Nine years ago the drought began—the worst in 120 years of record keeping. Adding to the problem is a rise in temperature of 1 degree Celsius (1.8 degrees Fahrenheit) over the last century, which has increased the rate of evaporation, especially in the summer. Farming activities have been dramatically curtailed, farm families are leaving their lands, and the water wars have begun pitting states against states, cities against rural residents, and environmentalists against irrigation.

Desertification seemed to be well under way in the Murray-Darling Basin. And then came the deluge. In the first three months of 2010 rains of a magnitude not seen for decades filled the rivers, reservoirs, and lakes all over New South Wales. Immediately politics came to the fore in the battle between those who wished to go back to the good old days of water for all and those who recognized that this was likely a brief respite in a long-term problem of controlling a vanishing resource. The *Economist* phrased the dilemma this way: Should Australia save its rivers, or its farmers?

MURRAY RIVER, AUSTRALIA

The nine-year drought in Australia's Murray-Darling River Basin was the greatest drought in 120 years of record keeping. It ended temporarily with the return of normal rainfall in 2010. Widespread desertification in areas that already have low rainfall (often accelerated by overgrazing) is an expected impact of global warming in regions such as western North America, the coastal zone of Mozambique, and parts of China.

Floods

Increased atmospheric temperatures will lead to increased water content in the atmosphere. For every increase of 1 degree Celsius (1.8 degrees Fahrenheit), the atmosphere can hold about 7% more water. As winters warm, more of the precipitation will be in the form of rain rather than snow, leading to different runoff patterns, another factor that will undoubtedly contribute to more frequent flooding. Changes in climate will be quite complex, and as noted earlier, they are very difficult if not impossible to predict on a local scale. Some changes in rainfall patterns will be seasonal. For example, Germany is expected to be drier in the summer and wetter in the winter as global change proceeds.

More important perhaps than average and seasonal rainfall changes is the likelihood of more extreme rainfall events, which will lead to more flooding. The fourteen inches of rain that fell on the first weekend of May 2010 in Nashville is a good example of an extreme event. The frequency of intense bursts of rain will be higher, even in arid areas.

For the insurance industry the importance of increased flooding is clear. Industry representatives note that flooding accounts for about one-tenth of all property insurance payouts, one-third of all economic losses from natural catastrophes, and more than half of all deaths from such catastrophes. The costliest insurance loss ever was incurred as a result of the Mississippi River flood of 1993.

Deaths in North America from river floods have dramatically decreased in the past few years because of improved warning systems. This is not so for deaths from hurricane storm surge flooding and tsunami flooding, as seen in Asia recently. Cyclone Nargis killed 133,000 people in Myanmar in 2008, and the great Indian Ocean tsunami killed 200,000 people in 2004. The massive Indus River floods that hit Pakistan in 2010 (said to be the worst in a thousand years) inundated about 20% of the country, destroying over a million homes and leaving over twenty million people homeless—astounding numbers.

Areas likely to be at risk from increased floods are the American Southwest (flash floods), the Northeast (spring floods), and the Northwest (more fall and winter flooding). Northern to northeastern Europe will also be at increased flood risk.

Admittedly, the warming atmosphere is not the only cause of increased flood damage. Other causes include increases in human population (which means there is more property to damage), in the value of property, in the amount of construction in flood-prone areas, and in the area of impermeable surfaces (e.g., concrete), as well as misplaced trust in engineering systems (e.g., levees).

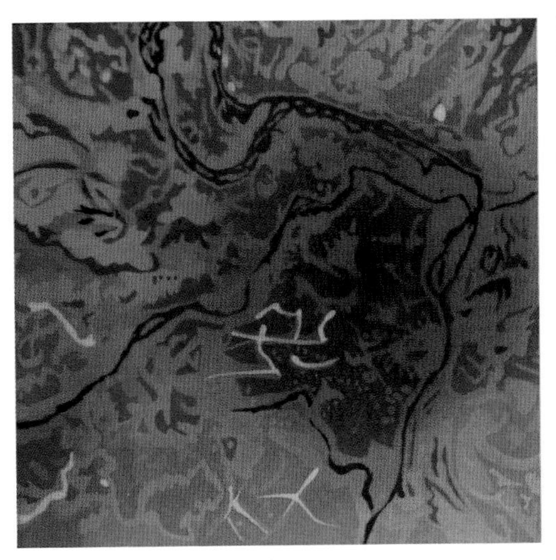

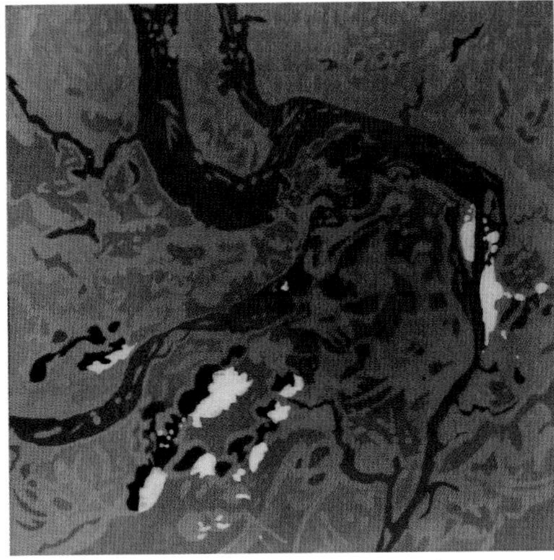

ST. LOUIS, 1988 AND 1993

Shown here are images of the junction of the Mississippi and Missouri rivers near St. Louis. To the left is the normal stage of the rivers in 1988; to the right, a view of the same area during the flood of 1993. Increased extreme rainfall events (for example the deluge in Nashville in May 2010 and the floods in North Korea in the summer of 2005) are likely to become more frequent because of global change.

Storms

There is considerable speculation over whether global change, especially the warming ocean, will lead to more frequent storms and more intense storms. Foremost among the storms that will strike and damage the North American mainland are the tropical storms, the largest of which are hurricanes. Other storms called winter storms—Sou'westers on the Pacific Coast and Nor'easters on the East Coast—also can be very damaging. Although their intensity may be below that of a full hurricane, because they often move slowly across the shoreline, rather than quickly pass over it as hurricanes usually do, these storms can create large storm surges and threaten much property. The Ash Wednesday storm that struck the East Coast in 1962 and caused serious damage from Massachusetts to north Florida may have been the region's most damaging storm of the twentieth century.

Tropical storms form over warm tropical waters when the temperature of surface waters is at least 27 degrees Celsius (80 degrees Fahrenheit). Under these warm conditions evaporation generates high humidity and clouds that lead to thunderstorms,

HURRICANE KATRINA

High-wind storms, such as hurricanes, cyclones, and typhoons, are predicted to occur with greater intensity and possibly greater frequency in coming decades. Property damage will increase accordingly, especially as the global population grows and more people crowd the hazardous areas in the world's coastal zones.

which may converge and begin to rotate, forming a tropical depression. When the winds exceed 56 kilometers (35 miles) an hour the system is called a tropical storm. A hurricane (or typhoon or cyclone in other parts of the world) is an intense tropical storm with winds over 119 kilometers (74 miles) an hour.

Since such storms gain their energy from the warm surface waters of the ocean, it seems reasonable to assume that as ocean waters warm, these tropical storms will increase in intensity. If they do, we should expect future hurricanes to have, on average, higher peak winds and greater volumes of rainfall. However, there is another possible effect of global warming that will determine whether there will be more and stronger hurricanes. Stronger heating of the Earth's surface will amplify the difference in temperature between the troposphere and the stratosphere, causing the two layers to move about in different directions and at different speeds. Their differing patterns will have a tendency to chop off the tops of tropical storms and reduce their intensity.

Just like the increase in global atmospheric temperatures, the change in storm activity is probably best evaluated over decades or even longer periods rather than year to year. The point is that the extent and intensity of storm activity is a very "noisy" curve, meaning that there is much variability from one year to the next. The available North Atlantic hurricane data for the decades 1945 to 1955 and 1995 to 2005 indicate a significant difference in hurricane character. For example, in the decade 1945–55 there were 74 hurricanes, 19 of which were category 4 or 5. Four decades later there were 112 hurricanes, 28 of which were category 4 or 5.

Globally there has been no trend, decadal or annual, toward increasingly frequent large tropical storms. But in the north Atlantic and Indian oceans the preliminary data, as just described above, indicate a possible trend of increasing tropical storm intensity. J. A. Curray and associates, in analyzing storm trends, believe that "time will tell" if hurricane intensity and frequency will increase. They argue that at least another decade of storm data is needed to see if a change is occurring.

It should be noted that the most gigantic of floods, the greatest of earthquakes, the largest of storms would be no catastrophe at all if humans weren't present. These events would just be natural curiosities. But the Earth's ever-increasing population assures that natural catastrophes will continue and increase over time. Andy Revkin of the *New York Times* notes that we face two climate threats. One is the threat from increased population, putting ever more people in the way of calamitous weather. The second is the threat that we create by producing greenhouse gases, possibly "making today's weather extremes tomorrow's norms" (*New York Times*, 8 September 2010).

GULF OIL SPILL

The BP Deep Horizon blow-out in the Gulf of Mexico in 2010 provides a strong reminder of how short-term human activities interact with longer-term global changes. Oil spills hasten the demise of coral reefs, salt marshes, and mangroves already under pressure from warming ocean waters and rising sea level.

Myths, Misinterpretations, and Misunderstandings of the Deniers

MYTH: *In the 1970s scientists were worried that the Earth was cooling, and now they've reversed their thinking. Global warming is just the new fashion.* A review of the scientific literature by Thomas Peterson and associates in 2008 showed not only that there was no rush by scientists in the 1970s to claim that the Earth was cooling, but in fact at that time human-caused warming dominated the peer-reviewed literature. The concern about a cooling Earth seems to have come from the media: "a major cooling widely considered to be inevitable" (*New York Times*, 21 May 1975); meteorologists are "almost unanimous" in belief in global cooling (*Newsweek*, 28 April 1975); "Telltale signs [of global cooling] are everywhere" (*Time*, 24 June 1974).

MYTH: *Since 1998 the Earth's atmosphere has begun to cool. So global warming is over.* The origin of this claim is that the global average temperature in 1998 was the highest recorded to date. Starting from this high point in a curve with large ups and downs, some deniers have determined that the Earth is cooling. Determining trends by starting with high points (or low points) of a "noisy" curve is poor science. Noisy trends must be evaluated over a long period, not year by year. In November 2009 the Natural Environment Research Council and the Royal Society in London issued a statement that the previous ten years had been the hottest on record. The warmest years on record, since accurate measurements began in the late 1800s, were 1934, 1998, and 2005.

MYTH: *It was hotter in the "Medieval Warm Period" than at present, so the current warming trend is just another blip in atmospheric temperatures.* It is true that between AD 800 and 1300 Europe and Eastern North America warmed up, but this warming, like the cooling of the Little Ice Age (1550 to 1850), may have been a regional climate change without global implications. An overwhelming number of climate scientists believe that the current warming is not just another blip. We are causing this one, and we are here to suffer the consequences. In the past there were probably many thousands of equivalents to the Medieval Warm Period and the Little Ice Age. Apparently neither the Medieval Warm Period nor the Little Ice Age affected sea level in any substantial way. If the cooling or warming had been global, the sea level should have dropped or risen in response to changes in ice accumulation in Greenland and the Antarctic, just as is happening today. In fact, the evidence (from ice cores) indicates that current temperatures may already be higher than those in the Medieval Warm Period.

MYTH: *Variations in solar activity are responsible for global warming. Stop worrying about greenhouse gas emissions and stop picking on coal and oil companies.* Variations in the Earth's orbit around the sun, and hence variations in solar input, were the major cause of global climate change in the eons of geologic time that preceded the last thirty years. The physicist Ilya Usoskin showed that before 1975 a strong relationship between global warming and changes in solar radiation existed for more than a thousand years. However, over the last thirty years "the climate and solar data diverge strongly from each other." In fact over the last thirty years solar activity has been at a minimum. In other words, because of the greenhouse effect we are out of step with the natural cycle of solar input to climate, and this should indeed worry us.

MYTH: *The hockey stick temperature curve doesn't exist.* The thousand-year temperature curve proposed by Michael Mann and made famous by Al Gore was shaped like a horizontal hockey stick with the upturned, business end of the stick representing temperature changes in the northern hemisphere during the last hundred years. This image has proved a lightning rod for global warming skeptics, because the shape of the hockey stick becomes less clear depending on the dataset and statistical methods used, and the estimated error bars (showing the range of values within which the answer should fall) are large. Mann's original hockey stick was inspired by tree-ring analyses, based on the assumption that the nature of the rings reflects growing conditions. The analyses turned out to be faulty and resulted in a too-smooth curve. Nonetheless, many new temperature reconstructions have been made, and they all show temperature increases in the twentieth century, especially since 1970, in much the same pattern as the original hockey stick did. The difference is that the original hockey-stick shape has been somewhat blurred. So the bottom line remains the same: if we look back at the last millennium, we see a significant rise in temperatures in the twentieth century. The hockey stick limps along.

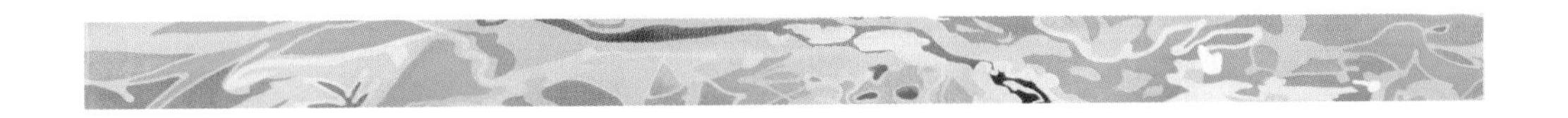

DOUBTS, UNCERTAINTIES, AND QUALMS ✳ 3

Global Perception

There are a number of important uncertainties in the science of global change. The Intergovernmental Panel on Climate Change (IPCC) is painfully aware of them and lists them in its various publications, especially in those published by the United Nations Environmental Programme (UNEP) publications, written for laypersons and available from www.earthprint.com. These uncertainties are not the Achilles' heels of global change. "It's almost certain that you can't put a trillion tons of CO_2 into the atmosphere without something nasty happening." This is according to James Lovelock, environmental thinker and futurist. Global change is a certainty and the direction of change is well understood. But rates, volumes, and levels remain uncertain.

There are a number of examples of well-founded and scientifically reasonable, peer-reviewed studies that question some of the numbers floating around the global climate change realm. None that we are aware of, however, question the reality of global warming and the human connection. In coming years some of the principles now assumed to govern climate change will themselves change as our science advances. It is the normal scheme of things for ideas about nature to change as we observe the workings of the natural world.

Most fundamentally, scientists are not exactly sure what will happen when we reduce greenhouse gas (GHG) emissions. What level of GHG emissions is required to halt global warming or to bring CO_2 concentrations in the atmosphere to a particular level? If we succeed in reducing CO_2, how fast will global changes already occurring reverse themselves? For example, according to the IPCC, sea level rise caused by the heating of ocean water will continue for centuries even after CO_2 levels have been stabilized.

Roger Pielke Jr. of the environmental studies program at the University of Colorado, Boulder, is one who argues that reducing greenhouse gases will not have a perceptible impact for decades. Although he is strongly in favor of reducing our greenhouse gas emissions, Pielke argues that the only effective response to global warming is adapting to new conditions rather than

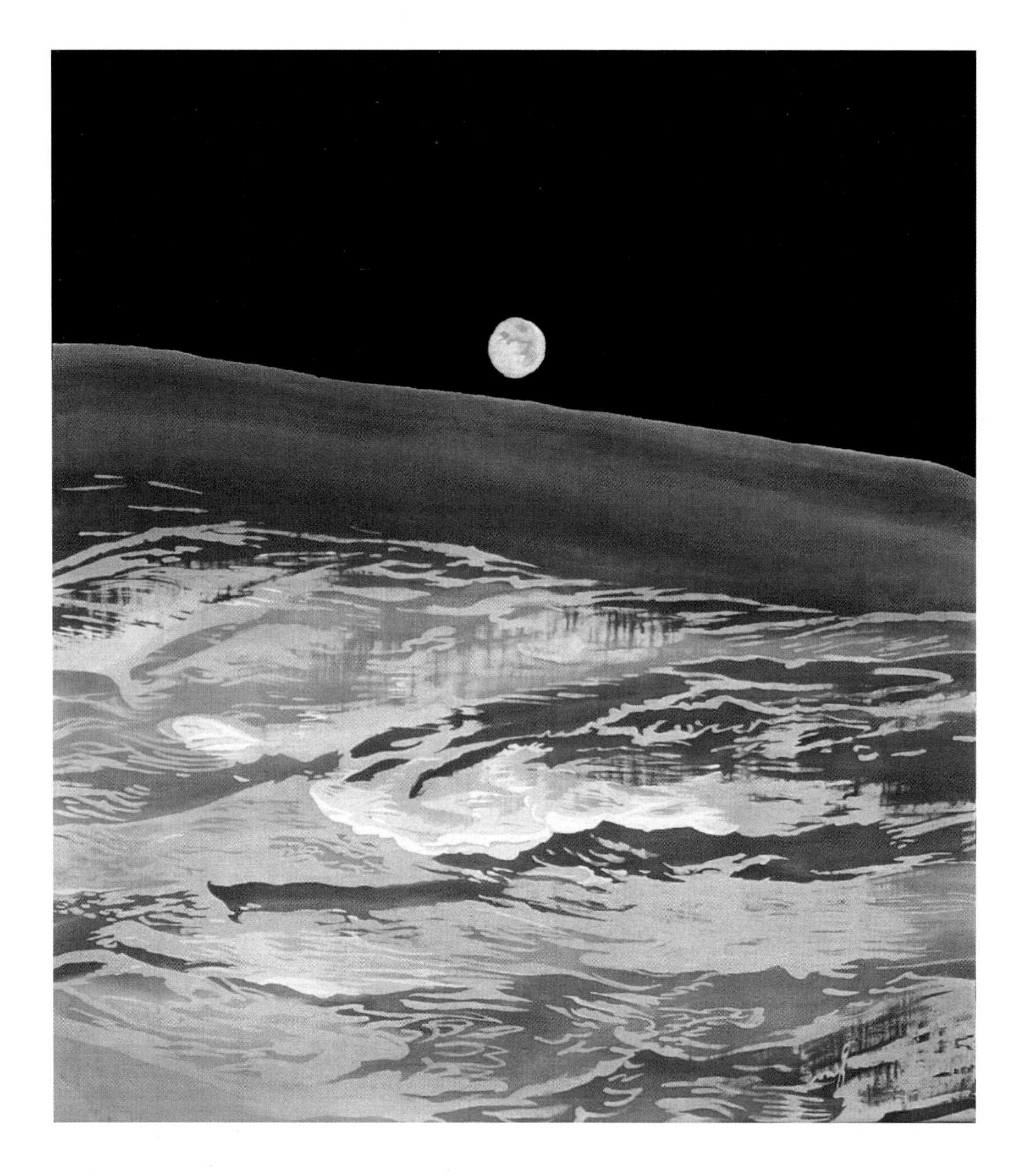

PACIFIC FULL MOON

The rise in sea level may well be the first catastrophe of truly global proportions caused by the increase in greenhouse gases in the atmosphere. A sea level rise of one meter (about three feet) is anticipated by the year 2100, but within forty to sixty years, millions of environmental refugees will be displaced; coastal cities, especially low-lying ones like Miami and Boston, will spend huge sums to preserve their infrastructure and buildings; and armed conflicts over water and territory can be expected.

waiting for conditions to change. In his book *The Climate Fix* (2010), Pielke concludes that decarbonization (lowering CO_2 production) won't work, because as long as global economies may be at stake, effective inaction on CO_2 reduction by the world's nations is almost a certainty. His solution: alternative sources of energy.

The scientific community is also not exactly sure what will happen as we continue to increase greenhouse gases in the atmosphere. The amount of future warming of the atmosphere per unit volume of released GHG is uncertain, because there are unknowns such as the proportions of different gases, the role of cloud cover and aerosols, the nature of future solar radiation, and the role of poorly understood feedbacks. Are there tipping points out there from which we cannot recover? We have some general understanding of these things, but our knowledge is certainly not precise. And it probably never will be precise, because nature defies complete understanding. This problem often frustrates policymakers, and provides ammunition for skeptics who often throw out the baby of well-understood phenomena with the bathwater of less-well-understood phenomena.

The problem of how climate change will play out and what causes it on a continental and global scale is generally understood, but at a local level climate change remains a major uncertainty. At a local level, events unrelated to global climate change such as pollution, industrialization, dam construction, and deforestation may be responsible for some changes. In the United Kingdom scientists provided regional projections of climate change using a grid made up of 25-km^2 (15.5 square-mile) squares. The report was supposed to be released in 2008 but it was not, because an independent review concluded that the limitations, which were quite serious, were not made clear in the report.

A number of the Earth and atmospheric processes that cause global temperature change are not well understood. These include the roles of aerosols and cloud cover, the future of ocean warming, and the various components of the earth's carbon cycle. In addition, the scale of methane emissions that can be expected from various sources and the scale of CO_2 emanating as a result of land use changes remain uncertain.

Extreme events, including storms, tropical hurricanes, intense precipitation (rain, sleet, snow), extreme temperatures, and droughts, are very important elements of global climate change and are important to forecast. But because these are uncommon events, they are harder to forecast and characterize than climatic averages.

Future changes in the ice sheets will have a major impact on sea level rise and are not easily predictable. The West Antarctic ice sheet is a particular enigma. Its melting rate may depend on the survival of ice shelves that hold back the outlet glaciers, the release from grounding on continental shelf islands as glaciers melt and become lighter, and warm currents that have begun, just in the last decade, to melt ice at the ice sheet margin. In addition there are new concerns about melting of the East Antarctic Ice Sheet (chapter 4).

Natural solar variability related to changes in the Earth's orbit is not understood well enough to accurately predict long-term natural changes that will underlie future global changes caused by humans. This is the conclusion of a study from 2010 headed up by the climatologist Eelco Rohling. Since the role of solar variability is imperfectly known, the projection of future climate over a century or longer is made more difficult. It is important to note that some scientists, particularly Wallace Broecker of Lamont-Doherty Earth Observatory, argue based on past cycles that we are on the verge of entering a new ice age. If true, the timing of this climate reversal is unknown, and some argue that warming caused by human beings is counteracting any cooling trends.

The Lightning Rods

CLIMATEGATE

In November 2009 the bombshell struck. More than a thousand e-mails were stolen from the Climate Research Unit of the University of East Anglia in Norwich. The e-mails were a bonanza for deniers who claimed that they represented a conspiracy to promote global change. They shed a bad light on the ethics of some climate scientists from East Anglia and their correspondents. The most damning discussions in the e-mails included apparent threats to obstruct the peer-review system to block publication of unwanted papers. It was unquestionably an example of scientists venting frustrations and perhaps of scientific arrogance, but there was certainly more smoke than fire. As the *Guardian* (29 December 2009) aptly characterized it, the e-mails show sincere researchers struggling to do good work in a highly politicized environment and sometimes losing their tempers. Climategate is perhaps best viewed as another example of the lengths to which climate deniers backed by corporate interests will stoop—in this case thievery—to promote their anti-scientific, pro-industry agenda.

Patrick Michaels, a senior fellow of the Cato Institute, one of the dozens of corporate-sponsored organizations dedicated to the manufacture of doubt, has written that the e-mails explain why a number of atmospheric scientists whose views differ from the majority have had difficulty publishing in top climate journals. There is no evidence that any of the e-mails in question actually resulted in the suppression of science. The examples that Michaels noted of scientists who are having alleged problems in publishing included himself, Willie Soon, Roy Spencer, and Sallie Baliunas—all relentless deniers who have made unsupportable statements against global warming in the past and who pursue their trade of global change skepticism with blinders that seem to prevent a balanced view of the science.

The peer-review process is a time-honored practice integral to the scientific method. Articles submitted to journals are anonymously reviewed by a group of peer experts in the field to insure that the articles are of high quality, based on sound scholarship and accurate research. A peer-review system that keeps out the publications of Patrick Michaels, who acts more like a lobbyist than a scientist, must be a good one! Michaels and others like him are not having difficulty publishing articles because of their minority views. Rather, the peer-review process reasonably prevents the publication of papers that are more propaganda than good science.

The problem with Michaels and others like him is akin to the problem that Earth scientists have with creationists. The Geological Society of America, for instance, receives requests each year from creationists asking to present technical papers at national meetings. Whenever creationists have been allowed to present their views at geological conferences, their very presence at a conference was later ballyhooed by their supporters as evidence that creationism was being accepted by a geological society. But when a geological society refuses to allow creationists to participate in its activities, the refusal is cited as evidence that the scientific mainstream has something to fear.

One of the most damaging leaked e-mails from Climategate reportedly came from Phil Jones, head of the climate change unit at East Anglia, who wrote, "I can't see either of these papers being in the next IPCC report. Kevin and I will keep them out somehow—even if we have to redefine what the peer reviewed literature is!" As Jones's statement shows, he was frustrated with the publication of even weak papers helping to legitimize climate change deniers, but his promise to redefine the peer-review process was certainly hyperbolic: no one person could hope to achieve that. George Monbiot of the *Guardian* noted that one of the papers in question was eventually published in the journal *Climate Research*, but it was so flawed that the resulting scandal led to the resignation of the editor in chief.

Although no real evidence of a grand conspiracy to distort or manipulate data was revealed by Climategate, the incident was nonetheless labeled as "an arrow through the heart of climate science," "the final nail in the coffin of anthropogenic warming," and much more by global warming deniers. Patrick Michaels pronounced that the "refereed literature has been inestimably damaged." The *Wall Street Journal* carried the hysteria even further by proclaiming, in an incredible statement, that Climategate had "harsh implications for the credibility of science generally." But a society that tramples on science does so at a huge risk to its own well-being, and if all American science is suspect, as the *Journal* claims, we are indeed in trouble. The one-time governor turned Fox News commentator

Sarah Palin opined that if President Obama stayed home from the Copenhagen Conference in 2009 he would have sent the message that "the United States will not be a party to fraudulent scientific practices."

As it turns out, a parliamentary committee in Britain cleared the Climategate scientists of all but the charge of unwillingness to share data; three other panels at the University of Virginia and Penn State eventually cleared them as well. Although it is hard to deny the arrogance of the scientists involved and their contempt for the norms of good scientific practice, it turns out that Climategate was largely a hyped-up media phenomenon. Mostly the scientists involved were grousing and griping, nothing more. Nonetheless, the attorney general of Virginia, who believes that global change is a myth, is seeking to obtain the research papers and field data from Michael Mann, one of the Climategate scientists.

Climategate is an example of the pot calling the kettle black. Many of the arguments from climate denialist groups are so blatantly false and deceitful that the hullabaloo over Climategate is hypocritical in the extreme. The list of their disingenuous claims (some listed at the end of each chapter) gives them no credible basis to accuse others of being disingenuous. In short, there is no conspiracy to exaggerate or manipulate data or computerized climate models. Even if an individual instance of manipulation were to come to light, it would not change the scientific facts determined by the community of global climate change researchers. Sadly, this will not prevent global warming deniers from trying to exploit these stolen e-mails to further their own agenda.

THE INTERGOVERNMENTAL PANEL ON CLIMATE CHANGE

The IPCC is the main lightning rod for much of the criticism by the deniers. Indeed they have made errors, as would any panel of more than two thousand scientists speaking many languages and meeting in a politically charged atmosphere. Our opinion is that the IPCC, while not perfect, is a group of largely honest, dedicated researchers working for little compensation under difficult conditions. That a committee of thousands could agree upon and produce massive and detailed reports is something of a miracle.

Mistakes are inevitably made. One serious error (called Glaciergate by some) was the claim that the Himalayan glaciers would disappear by 2035. It apparently began with a misquote by the World Wildlife Fund in one of its reports. The error was immediately trumpeted by critics as another example of IPCC's incompetence and tendency to exaggerate, and a reason to question the credibility of all the panel's work. This is another good example of throwing out the baby with the bathwater.

Glaciergate won't be the last mistake that will be discovered. Unfortunately each error, and even each change in forecasts based on advances in understanding of

ICEBERG

This iceberg batik shows that most of the ice is below the surface. As the world's ice sheets and mountain glaciers melt, floating icebergs calved from the noses of the glaciers that extend into the sea will become more common. To determine the net contribution of water to the sea from melting and calving glaciers, it is necessary to know the contribution of new snow added to the interior each winter.

various global change processes, will likely be falsely characterized as further evidence of the IPCC's ineptitude and even fraud.

Elizabeth Kolbert writes: "No one has ever offered a plausible account of why thousands of scientists at hundreds of universities in dozens of countries would bother to engineer a climate hoax. Nor has anyone been able to explain why Mother Nature would keep playing along; despite what it might have felt like in the northeast these past few months, globally it was one of the warmest winters on record."

In IPCC's 4th Assessment Report (2007), or AR4, the panel committed an editorial error by stating that sea level would rise between seven and twenty-three inches by 2100, *not counting the contribution of the melting ice sheets*. The absence of the ice sheet contribution was duly but not clearly noted in the text of the report. Some who read this report misinterpreted the rise of seven to twenty-three inches as being the expected total rise. Others among the deniers have chosen to use these numbers to minimize the potential importance of sea level rise. The failure of the IPCC to include the ice sheet contribution in its forecast was regrettable, since clearly the melting of ice sheets will have a significant impact on sea level rise.

MATHEMATICAL MODELS

Much praise and criticism is showered upon mathematical models (computer models). Accurate prediction of future trends in atmospheric warming, permafrost melting, ice sheet degradation, and sea level rise is a perilous process. Extensive field observations must provide the main basis for projections, but beyond fifty years out, mathematical models prevail. It is our belief that scientists sometimes place too much faith in mathematical models.

There are two kinds of mathematical models, quantitative and qualitative. Quantitative models provide a number by asking where, when, and how much. Qualitative models provide an understanding by asking how, why, and what if. Global change models, taken qualitatively, have provided a critical understanding of the process of change, have given estimates of direction and orders of magnitude, and have answered numerous "what if" questions. Taken quantitatively, models are another matter. As the discussion above indicates, there are a large number of uncertainties, certainly enough to make anyone nervous about the various numbers associated with the global warming phenomenon. For example, James Hansen's claim that an atmospheric CO_2 concentration of 350 parts per million marks an irreversible global change tipping point is most difficult to defend. A tipping point may well exist, but a precise number is simply not justifiable in the context of so many uncertainties.

Another example of overreaching in mathematical models was the statement made by global change scientists testifying before Congress in 2007. They observed that according to models, we have ten to twelve years to start reversing the rate of greenhouse gas emissions before the global changes become irreversible. Again the uncertainties make such an assertion very tenuous.

A different type of problem with mathematical modeling is illustrated by the IPCC's aforementioned failure to include melting ice sheets in its projection of sea level change by the turn of this century. This failure occurred because the panel said there were too many uncertainties to make a projection. It was a case of overdependence on models: since the panel couldn't model the effect of melting ice sheets, it would not make any projection at all. Others, mostly on state and national sea level rise panels, have had to roughly estimate the contribution of the ice sheets based on field and satellite observations.

Claims that a model is valid because it has successfully reproduced a past event (hindcasting) should also be viewed with skepticism. As pointed out by Naomi Oreskes of the University of California, San Diego, hindcasting assumes that the complex natural systems being modeled will be subject to the same forces and will respond the same way all the time—past, present, and future.

For example, the model GENESIS is used by coastal engineers to predict, among other things, the amount of beach erosion in the future. The model is "calibrated" by hindcasting the erosion during some previous time span. The model is "adjusted" so it comes up with the right answer and then is applied to the future. But among other problems, beach erosion is very sensitive to storms; their intensity, duration, and the direction from which they come are variable factors that contribute to changing erosion patterns. To apply the model to the future requires the unlikely assumption that the schedule of storms in the future will be quite similar to the schedule of storms in the past.

With regard to predictive mathematical modeling, the ever-colorful James Lovelock argues, "we tend to be too hubristic to notice the limitations. If you make a model, after a while you get suckered into it. You begin to forget that it's a model and think of it as the real world. You really start to believe it. We really don't know what the clouds and aerosols are doing. They could be absolutely running the show." As a result of our belief in models we are minimizing direct observational data.

Jumping on Bandwagons

Research scientists tend to be a skeptical lot, but a couple of human-nature hazards facing them can dull the sharp edge of scientific cynicism that characterizes good science. One of these hazards is the "bandwagon" effect and the other is the "state of siege" effect. The bandwagon effect occurs when the vast majority of scientists support some idea such as some element of global change. Their support for the idea can cause the less venturesome and less courageous in the scientific community to jump up on the bandwagon and support the idea unquestioningly just to take advantage of safety in numbers. Global change science is a bandwagon if ever there was one.

The state of siege that climate scientists find themselves in today has a similar dampening effect, because of the verbal abuse that inevitably follows the announcement of a new scientific observation that favors the human connection to climate change. This problem is particularly real for scientists who interact with the public through media interviews, newspaper articles, speeches, and seminars. Why ask for trouble? Some find it better to remain in the background.

Is Carbon Dioxide Good for the Planet?

Among the global warming contrarians is a particular breed of CO_2 "boosters." These CO_2 cheerleaders argue that more carbon dioxide in the atmosphere and oceans will only increase plant and animal life. Chief among these proponents is the Idso clan, Sherwood Idso and his sons Craig and Keith Idso. Sherwood Idso was affiliated with the Greening Earth Society, a group formed and funded by Western Fuels Association to promote the idea that increasing CO_2 is good for humanity. They produced two documentaries, the first of which, *The Greening of Planet Earth: The Effects of Carbon Dioxide on the Biosphere*, argues that modern agricultural advances are attributable in part to increased carbon dioxide in the atmosphere, thanks to the Industrial Revolution. The Idso brothers lead the Center for the Study of Carbon Dioxide and Global Change, which was "created to disseminate factual reports and sound commentary on new developments in the worldwide scientific quest to determine the climatic and biological consequences of the ongoing rise in the air's CO_2 content."

The geologist H. Leighton Steward, in his self-published book *Fire, Ice, and Paradise*, suggests that the "old wives' tale" that plants grow better if you talk to them might be related to the CO_2 which is released with every exhale of breath. Steward rejects the notion that CO_2 should be considered a pollutant, but instead considers it a "great airborne fertilizer which, as its concentrations rise, causes additional plant growth and causes plants to need less water." Steward also attributes 12% to 15% of worldwide plant growth to the "100 ppm of CO_2 added to the atmosphere since the start of the industrial revolution." Steward maintains that the Earth needs more CO_2, not less, a view consistent with the aforementioned documentary *The Greening of Planet Earth*, which proclaims, "more scientists are confirming our world is deficient in CO_2 and a doubling of atmospheric CO_2 is very beneficial."

The potential positive impact on agriculture is a commonly cited positive outcome of increased CO_2 in the atmosphere. Studies have shown that CO_2 can boost plant productivity and some greenhouse owners actually purchase CO_2 to increase plant growth.

In an exchange before the House Subcommittee on Energy and Environment in March 2009, the British climate change contrarian Christopher Walter Monckton, 3rd Viscount Monckton of Brenchley, argued that "carbon dioxide is a plant food," which prompted Representative John Shimkus to ask: "So if we decrease the use of carbon dioxide, are we not taking away plant food from the atmosphere?"

As an elementary school student, you most likely learned that plants use CO_2 during photosynthesis. Increased CO_2 has been shown to accelerate the rate of photosynthesis. Simply put, the pro-CO_2 argument is that since plants need CO_2, they will benefit from increased levels of it in the atmosphere. This is essentially the argument being put forward by CO_2 boosters, but it is a gross oversimplification. For one thing, anticipated climate change does not involve only increased CO_2, and studies have shown that increased CO_2 may actually negate some of the agricultural benefits of global warming.

M. Rebecca Shaw, a botanist at the Carnegie Institution, and her colleagues found that anticipated climate changes of warming, increased precipitation, and nitrogen deposition, alone or in combination, increased net primary production in the third year of ecosystem-scale manipulations in annual grasslands in California. While increased CO_2 by itself also improved net primary production, "across all multifactor manipulations, elevated carbon dioxide suppressed root allocation, decreasing the positive effects of increased temperature, precipitation, and nitrogen deposition on net primary production."

Further, an article by the freelance writer Ned Stafford in the journal *Nature* in 2007 cited several studies suggesting that more carbon dioxide will result in less nutrition in plants, in part because of decreased intake of nitrogen, calcium, and zinc. Clenton Owensby and other researchers at Kansas State also found that elevated CO_2 levels resulted in less nutritious and less digestible grass for cattle, suggesting that future ruminants may gain less weight even if they eat more grass in a CO_2-enriched world. They also noted that insects appear to increase consumption as nutrition decreases in an elevated CO_2 setting.

The Argentine scientist Jorge Zavala performed an open field study involving the levels of CO_2 expected to be in the atmosphere by the year 2050 and found that soybeans produced less jasmonic acid, a natural defense to insect pests, allowing adult insects to feast on the plants, live longer, and produce more offspring.

Increased levels of carbon dioxide may contribute to warming in another manner apart from the greenhouse effect. Long Cao and Ken Caldeira, researchers at the Carnegie Institution, have found that increased CO_2 negates the cooling effect of trees and contributes to warming. Plants give off water through tiny pores in their leaves in a process called evapotranspiration. This process not only cools the trees but also releases water into the air, cooling the surroundings. Increased carbon dioxide causes the leaves' pores to shrink and release less water. Cao and Caldeira found that in some regions, including North America and East Asia, over a quarter of warming from increased CO_2 in the atmosphere is a result of decreased evaporative cooling by plants. They also found that high carbon dioxide will result in greater runoff from the land surface, as water from precipitation increasingly bypasses the plant cooling system and flows directly into rivers and streams.

Thus the answer to the question of how valuable increased CO_2 is to the kingdom of plants is not a simple one. Yes, by some measures plant growth is enhanced, but a number of other, less desirable effects occur as well. Once again there is proof that nature is never simple. But to look at CO_2 and plant interaction to the exclusion of other impacts of global change such as sea level rise is both absurd and irresponsible.

Myths, Misinterpretations, and Misunderstandings of the Deniers

MYTH: *The IPCC's mistakes tend to exaggerate or overblow the extent of climate change, thereby demonstrating a systematic bias.* Far from it; we believe that the IPCC tends to be conservative, probably on purpose. The deniers jump exclusively on mistakes that exaggerate climate change while ignoring mistakes that underestimate it. A case in point is sea level rise. The IPCC has been consistently low in its estimates of sea level rise rates in its last two reports. In 2001 it incorrectly assumed that the Antarctic ice sheet was not going to contribute melt water to the rising sea level, a mistake that it corrected in 2007.

Bjørn Lomborg, economist and adjunct professor at Copenhagen Business School, is among the most sophisticated of the skeptics and in some circles is considered a legitimate critic, but he is not. Lomborg came on the scene in a big way with the publication of his best-selling book *The Skeptical Environmentalist* (2001), and in 2007 he published *Cool It: The Skeptical Environmentalist's Guide to Global Warming.* Both books deal with similar themes, mainly that claims of catastrophe are overblown and that cost-benefit analysis suggests the need to pay more attention to other problems, such as poverty, AIDS, and malaria. The most likely reason that the books have had such a large impact is that they have the appearance, if not the substance, of a legitimate, well-documented review of the science of global change.

Lomborg's review of the science is inaccurate, however, as documented in painful detail by Howard Friel in his book *The Lomborg Deception* (2010). Friel and others point out that Lomborg's apparent reliance on peer-reviewed scientific literature is highly misleading. Often the literature is misquoted or quoted out of context. It is fair to say that Lomborg's views are completely out of step with the views of global change scientists.

James Gustave Speth, dean of the Yale School of Forestry and Environmental Studies, notes that "for nearly a decade, Bjørn Lomborg's climate-science rejectionism has helped block serious political action on greenhouse emissions." In 2003 the Danish Committees on Scientific Dishonesty, organized under the nation's Ministry of Science, cited *The Skeptical Environmentalist* for data fabrication, cherry picking, plagiarism, deliberate misinterpretation, and use of misleading statistics. Below are paraphrases of some of Lomborg's statements about sea level rise.

MYTH *(straw man): Contrary to common statements, melting sea ice will not raise sea level.* No scientist that we know believes that melting sea ice (floating ice) on the surface of the sea will contribute much to sea level rise, and we have never seen this "common" statement. Thus Lomborg misleadingly rejects a proposition that is not made by scientists. Lomborg's focus on floating ice, rather than ice sheets, allows him to minimize the dangers of sea level rise.

MYTH *(misquotes): The IPCC report (2007) estimated that sea level will rise a foot in the twenty-first century.* True enough, but the report pointed out that the one foot mid-range number did not include ice sheet melting, which the IPCC expects may be the primary source of sea level rise later in this century. Thus Lomborg attempts to downplay the effect of sea level rise by ignoring the report as a whole.

MYTH *(cherry picking): Sea level will rise thirty centimeters.* Lomborg sticks with the projected rise in sea level of thirty centimeters (one foot) by 2100, although more than a dozen science panels around the globe have projected a minimal rise of one meter (three feet) by then. In addition, the current rate of sea level rise as measured by both satellites and tide gauges is greater than thirty centimeters (one foot) per century.

MYTH *(strange opinion). The concern for sea level rise is enhanced by society's biblical fear of flooding.* Lomborg's point is that environmentalists and scientists are sounding warnings about sea level rise more because of the ancient legends of floods than because of hard evidence of potential impacts. It is an absurdity.

The turnabout? In 2010 Lomborg published a book entitled *Smart Solutions to Climate Change*. In the pre-publication reviews he is said to argue that global warming is a major concern for the world and that $100 billion per year will be needed to fight climate change. This is a drop in the bucket of global change response. The book appears to be a major reversal of opinion on Lomborg's part, but in reality the low figure of $100 billion actually belittles the risks of global change.

THE MANUFACTURE OF DISSENT

4 ✱ THE GLOBAL WARMING DENIAL LOBBY

Senator James Inhofe (R-Okla.) calls the threat of catastrophic global warming "the greatest hoax ever perpetrated on the American people." It turns out that there is a hoax involving climate change. Only the hoax is being perpetrated by public relations efforts by the fossil fuels industry.

Tobacco Roots

Efforts by corporations to cast doubt on the veracity of global warming science are akin to the tobacco industry's campaign to cast doubt on the dangers of cigarettes. The links between the tobacco industry and global warming deniers were revealed in large part because of internal documents, which were posted to the Internet after a class-action suit against tobacco companies, as detailed in *Climate Cover-up* (2009), by James Hoggan and Richard Littlemore.

In response to an EPA report on the harmful effects of secondhand smoke, the tobacco giant Philip Morris hired APCO Worldwide, a public relations company. APCO created what is now known as an Astroturf group (that is, a fake "grassroots" group)—an entity known as The Advancement for Sound Science Coalition (TASSC). APCO's plan to create TASSC can be found online at TobaccoDocuments.org. In an effort to avoid being branded as an arm of Big Tobacco, TASSC sought to tackle "broader questions about government research and regulations," including global warming. In what would be established as a pattern, APCO launched a media campaign in regional markets outside large metropolitan areas where more aggressive or better-funded media might challenge its messages. Philip Morris sought to counteract science linking secondhand smoke to cancer and even appears to have coined the term "junk science" for peer-reviewed studies which might harm their industry, as opposed to "sound science," for studies which support their views. Of course Philip Morris was not alone in the quest to counter harmful scientific findings. A memo prepared at Brown and Williamson proclaimed: "Doubt is our product since it is the best means of competing with the 'body of fact' that exists in the mind of the general public. It is also the means of establishing a controversy."

In the report "Smoke, Mirrors and Hot Air" (2007), the Union of Concerned Scientists explored how ExxonMobil was employing Big Tobacco's tactics to manufacture uncertainty about global warming. The report identified five main tactics used by tobacco companies to cast doubt on the dangers of smoking: (1) questioning even indisputable scientific evidence showing their products to be hazardous; (2) engaging in "information laundering" by using and even establishing seemingly independent front organizations to make the industry's case and confuse the public; (3) promoting scientific spokespeople and investing in scientific research in an attempt to lend legitimacy to their public relations efforts; (4) attempting to recast the debate by charging that the wholly legitimate health concerns raised about smoking were not based upon "sound science"; and (5) cultivating close ties with government officials and members of Congress. The report shows how ExxonMobil uses these very same tactics to create doubt in the public's mind over the scientific consensus regarding global warming.

ExxonMobil is by no means the only petrochemical player to dabble in manufacturing doubt. In 1991 the Western Fuels Association teamed with the National Coal Association and the Edison Electric Institute to create a campaign to combat the public's growing concern over global warming. After considering such names as the Information Council for the Environment, Informed Citizens for the Environment, and Informed Choices for the Environment, they settled on the Information Council for the Environment (ICE). The science historian Naomi Oreskes has published their founding documents on the Internet. These documents show that ICE's strategy echoes the earlier campaigns by TASSC. ICE spelled out a seven-point strategy, beginning with an attempt to "reposition global warming as theory (not fact)." Another strategy was to "use a spokesman from the scientific community."

ICE chose four cities (Chattanooga, Tennessee; Champaign, Illinois; Flagstaff, Arizona; and Fargo, North Dakota), all of which were home to members of either the House Energy and Commerce Committee or the House Ways and Means Committee, as test markets for a $500,000 marketing campaign. The goals of the test markets were (1) to demonstrate that a consumer-based media awareness program can change the opinions of a selected population regarding the validity of global warming; (2) to begin developing a message and strategy for shaping public opinion on a national scale; and (3) to lay the groundwork for a unified front for the national electric industry on global warming.

The strategy included radio and print campaigns. The radio ads would "directly attack the proponents of global warming by relating irrefutable evidence to the contrary, delivered by a believable spokesperson," while the print ads would "attack proponents through comparison of global warming to historical or mythical instances of gloom and doom. Each ad will invite the listener/reader to call or write for further information, thus creating a data base." The test campaign reportedly revealed that audiences trusted "technical sources" most, activists and government officials next, and industry the least. ICE concluded that industry should find scientists to serve as spokesmen.

The campaign identified two target audiences: "Older, less educated males," who are receptive to "messages describing the motivations and vested interests of people currently making pronouncements on global warming—for example, the statement that some members of the media scare the public about global warming to increase their audience and their influence. . . ."; and "younger, lower-income women," who "are more receptive to factual information concerning the evidence for global warming. They are likely to be 'green' consumers, believe the earth is warming, and to think the problem is serious. However, they are also likely to soften their support for federal legislation after hearing new information."

Following the success of their test campaign, Western Fuels released the documentary *The Greening of Planet Earth: The Effects of Carbon Dioxide on the Biosphere*. As noted in chapter 3, the movie, produced by the Greening Earth Society and reportedly funded by Western Fuels, suggests that increased carbon dioxide would significantly benefit plant growth, which is an oversimplification. As of this writing, the video is available on YouTube. The now defunct Greening Earth Society also published a journal, edited by Patrick Michaels. In 1998 the Greening Earth Society produced a second documentary in which, according to the website of the Center for the Study of Carbon Dioxide and Global Change, "expert scientists assert that CO_2 is not a pollutant, but a nutrient to life on earth."

Delivering the Message: The Role of Media

Steven Milloy may be the most obvious example of a well-placed corporate spokesperson spreading the gospel of climate change denial. He got his start with Philip Morris and was once the executive director of the denial advocacy group The Advancement for Sound Science Coalition. Now he is a commentator for Fox News and also runs the website junkscience.com, a pro-industry website that includes articles denying or downplaying climate change. On his website Milloy, a self-proclaimed "Junkman," defines junk science as "faulty scientific data and analysis used to advance special and, often, hidden agendas." Milloy is certainly not a trained journalist: he is a hypocrite, guilty of the very tactics he claims to expose, using "faulty" science to advance his own hidden agenda.

The climate change denial industry need not rely on public relations firms and lobbyists posing as journalists. The very nature of the modern media plays a major role in promoting doubt over climate change.

When journalists, particularly television reporters, cover a story, they typically interview people representing opposing sides of an issue, in the climate change arena no less than in others. Thus despite the overwhelming scientific consensus that global warming is a serious, imminent danger, news stories in print or on television will inevitably feature someone with the "minority view," either denying or minimizing the risks of climate change. This keeps a handful of industry scientific spokespersons busy.

If you are cognizant, you will undoubtedly recognize the names and faces of these climate change deniers, as they frequently appear in the press. However, you needn't take our word for it. Jules and Max Boykoff compared the coverage of global warming in the scientific press to the coverage of the same issue in "prestige" United States newspapers from 1998 to 2002. They found that while there was vast agreement in the scientific community that "human actions are contributing to global warming," the majority of newspaper articles gave roughly equal treatment to the view that "humans were contributing to global warming, and the other view that exclusively natural fluctuations could explain the earth's temperature increase." The study concluded, "balanced reporting is actually problematic in practice when discussing the human contribution to global warming." "Balanced" reporting legitimizes marginalized views and contributes to the success of the climate change deniers.

Also contributing to the success of those who deny that climate change is a threat is the twenty-four-hour news cycle, which provides a forum to climate denial spokespersons (once again benefiting from the concept of "balanced" reporting). Conservative talk radio has also boosted the fortunes of the deniers by repeating their messages to an audience eager to receive anti-government, anti-scientific, and anti-"élitist" propaganda. And last but not least, the Internet and in particular the blogosphere provide endless opportunities to circulate articles or opinions long after they may have been debunked. Hoggan and Littlemore describe as an "echo chamber" the "reverberating network of think tanks, blogs, and ideologically sympathetic mainstream media outlets that distribute and circulate contrarian information."

Orrin Pilkey and Rob Young published an op-ed piece in *USA Today* in January 2010. They noted that there are many observable indicators of global warming besides atmospheric temperature measurements and climate models (no mention was made of the human connection). Within a couple of days there were about seven hundred replies from (we assume) nonscientists. What emerged from the response was that there is an element of the American population with an alarming distrust not only of their own government but also of science and scientists in general. What also emerged is a glimpse of the efficacy of the "echo chamber" and the success of the climate change denial industry. In rough order of frequency the responses were as follows:

- Global warming is unrelated to human activities.
- Global change is all part of natural Earth cycles.
- Crooked, lying scientists are behind global change (a clear response to Climategate).
- Crooked, lying politicians are behind this.
- Global warming is real but . . .
- There is no global warming.

Frequently the comments were quite harsh:

- The junk science you and your fellow falsifiers have foisted on our populace contains not a shred of credibility.
- We all know it's a huge scam worth billions of dollars, and that you are lying and concealing data, conspiring to shut out anyone who doesn't conform to your communist, one-world-government rule and wealth redistribution policies.

How effective is the denial echo chamber? A poll by the Pew Research Center in 2009 revealed a sharp decline in the percentage of Americans who say there is solid evidence that global temperatures are rising. In April 2008, 71% said there was solid evidence of rising global temperatures. Polling in September–October 2009 showed that just 57% of Americans felt that there was solid evidence of higher average temperatures over the past few decades. Also, the proportion of Americans who viewed global warming as a very serious problem dropped from 44% to 35% between April 2008 and September–October 2009. Meanwhile, a survey by Eurobarometer in 2009 indicated that Europeans view climate change as the second-most serious problem the world faces, behind "poverty, the lack of food and drinking water." Among those who do research in any aspect of climate change there is essentially no controversy concerning whether global warming is upon us or whether humans are at least partly the cause of the problem. There is a wide range of scientific opinion among researchers on the details of this global phenomenon. But let us not forget that the scientific debate is about details, not over whether global warming is real or whether it is related to human activities. Industry-backed think tanks and Astroturf organizations are simply taking advantage of the nature of scientific inquiry to manufacture doubt in the minds of the public and dissuade policymakers from taking action which might harm their interests. Scientists who are global warming skeptics or deniers have created the appearance of a controversy where none exists.

Peter T. Doran and Maggie Kendall Zimmerman of the University of Illinois, Chicago, surveyed 3,146 earth scientists (from a geoscience phone directory) and asked the following primary questions: (1) When compared with pre-1800s levels, do you think that mean global temperatures have generally risen, fallen, or remained relatively constant? (2) Do you think human activity is a

GLOBAL PERCEPTION

A global survey of the many physical changes that are occurring on the Earth's surface is verification that they are global in extent. Changes that are truly global include sea level rise, mountain glacier melting, and changes in plant and animal distribution. It is difficult to understand why deniers question global change, since these and other changes are easily measureable.

significant contributing factor in changing mean global temperatures? In response to the first question 90% of the respondents answered yes, as did 82% to the second question. The researchers found that in general, the percentages rise "as the level of active research and specialization in climate science increases," and that among those who cited climate science as their area of expertise and who had published more than 50% of their recent peer-reviewed papers on the subject of climate change, the proportion of positive responses rose to 96.2% for the first question and 97.4% for the second. Doran and Zimmerman concluded: "The debate on the authenticity of global warming and the role played by human activity is largely nonexistent among those who understand the nuances and scientific basis of long-term climate processes. The challenge, rather, appears to be how to effectively communicate this fact to policy makers and to a public that continues to mistakenly perceive a fundamental global change debate among scientists."

Scientists are certainly not infallible. Just like the rest of society they sometimes hop aboard bandwagons and interpret results according to prevailing popular views. But the very nature of science and the scientific method encourages skepticism. The global warming denial lobby takes advantage of this and promotes the views of an extreme minority to make it appear that there is no scientific consensus on global warming and the human role in climate change.

The Major Players (Myth Makers)

Among global change researchers there are essentially no global warming deniers, as already noted, but much discourse continues about the minutiae. Global warming deniers, on the other hand, have a different agenda: they seek a "truth" according to their clients' needs or according to their political beliefs. Those who seek to understand climate change as viewed by the scientific research community should ignore prominent spokespersons such as Patrick Michaels, Willie Soon, Nir Shaviv, Craig Idso, Richard Lindzen, Bjørn Lomborg, and S. Fred Singer. Organizations which appear dedicated to promoting skepticism on global climate change and to confusing the public and thus delaying action include the Cato Institute, the Heartland Institute, the American Enterprise Institute, the Competitive Enterprise Institute, and the George C. Marshall Foundation. If you read a statement about climate change with the name of any of the above-mentioned organizations attached, you should do so with the understanding that you are most likely reading global warming denier propaganda.

Not surprisingly, given the major role that fossil fuels play in the greenhouse effect, funding for climate change denier front groups and similarly minded "think tanks" frequently flows from the fossil fuels industry: the industries that stand to benefit if greenhouse gas emissions remain unchecked and that have the most to lose if governments are successful in promoting sustainable or alternative energy sources.

THE CARBON LOBBY

ExxonMobil, Western Fuel Industries, and the American Petroleum Institute are among the leading sources of funding for conservative think tanks that promote doubt over global warming science and oppose clean energy policy.

Between 1998 and 2005 ExxonMobil gave almost $16 million dollars to anti–global warming advocacy organizations. In 2008 the company publicly announced that it would stop funding anti–global warming organizations, but it still funds select groups. According to its own website, ExxonMobil continues to channel tens of thousands of dollars to groups such as the Heritage Foundation and the National Center for Policy Analysis, a group in Texas that according to its website believes the causes of warming are unknown and "the cost of actions to substantially reduce CO_2 emissions would be quite high and result in economic decline, accelerated environmental destruction, and do little or nothing to prevent global warming regardless of its cause." Further, industry-funded groups in the United States are spreading cash to support climate change skeptics in other countries. One reporter, Josh Harkinson, found that the Atlas Economic Research Foundation, a group based in the United States which received around $100,000 in 2008 from ExxonMobil, has supported more than thirty foreign think tanks "that espouse skepticism about the science of climate change." United States industry is exporting its brand of politicized corporate climate science. This is bad news, because global consensus and cooperation are essential in making the changes needed to slow the release of greenhouse gases that bring about climate change.

A report in March 2010 revealed that Koch Industries, one of the largest and wealthiest private corporations in America, is a leading contributor to global warming deniers and groups opposing clean energy reform, even outspending ExxonMobil: between 2005 and 2008 ExxonMobil spent $8.9 million, while foundations controlled by Koch Industries doled out $24.9 million to the climate denial lobby. Recipients of Koch money include the Heritage Foundation ($1,620,000), the Cato Institute ($1,028,400), and the Atlas Economic Research Foundation ($113,800).

Conservative, pro-business groups and fossil fuel industries are the main sources of funding for the parade of climate change deniers. The following is a sampling compiled by the Union of Concerned Scientists of funding sources for anti–global warming advocacy organizations. Each organization takes a different spin on climate change, but ultimately they all share the goal of delaying or defeating, by creating doubt, the major policy changes to combat climate change.

Myth Makers

Global Climate Coalition. *Funding:* forty-six corporations and trade associations. *Principal message:* Global warming is real but it is too costly to respond.

George Marshall Institute. *Funding:* ExxonMobil. *Principal message:* Variations in solar radiation cause warming.

Oregon Institute of Science and Medicine. *Funding:* Unknown private sources. *Principal message:* There is no global warming and the IPCC is a hoax.

Science and Environmental Policy Project. *Funding:* Conservative foundations and the Reverend Sun Myung Moon. *Principal message:* Climate change is good. Action is not warranted because of poor science.

Greening Earth Society. *Funding:* Western Fuels Association (coal and utility companies). *Principal message:* CO_2 is good for Earth and coal is the best energy source.

Center for the Study of Carbon Dioxide and Global Change. *Funding:* Probably Western Fuels. *Principal message:* Increased CO_2 will help plants.

The physicist Stefan Rahmstorf recognizes three subspecies among the naysayers: trend skeptics, attribution skeptics, and impact skeptics. Trend skeptics deny any trend of global warming altogether. Attribution skeptics agree that warming is occurring but argue that you can't attribute it to human activities. Impact skeptics agree that warming is occurring but say that its impact will be mostly positive.

There is another type of skeptic, which we might call the pessimistic believer. People in this category believe in the human connection and may even favor greenhouse gas reduction, but they also believe that action is futile or too expensive to pursue. Some of these skeptics may be viewed as delayers rather than deniers. If they can admit the legitimacy of the scientific consensus but cast doubt on whether action can be effective, they can delay any action and further benefit their constituents—the fossil fuel industry. They typically argue that action to combat global change will be economically damaging, which is a valid point, but one should recognize that it would be particularly damaging to energy giants such as ExxonMobil and Koch Industries.

Then there is the philosophical believer who doesn't deny global climate change but argues that we are worked up to an unjustifiable pitch of panic, fueled by environmental ideologues. In an op-ed piece published in the *New York Times* on 1 January 2010, the New Zealand philosopher Dennis Dutton compared climate "hysteria" to that surrounding the Y2K computer problem at the turn of the century: "Apocalyptic projections are a diversion from real problems—poverty, terrorism, broken financial systems . . . this applies, in my view, to the towering seas, storms, droughts, and mass extinctions of popular climate catastrophism. Such entertaining visions owe less to scientific climatology than to eschatology and that familiar sense that modernity and its wasteful comforts are bringing us closer to a Biblical day of judgement."

There is a certain irony to Dutton's likening of what he perceives as global warming hysteria to the notion that modernity is leading us to a "Biblical day of judgement," given that a large share of climate change naysayers are fundamental Christians of a conservative bent. It is true that the news media will often inflate a problem out of proportion to gain an audience. However, Dutton's view ignores the fact that much of the concern about climate change is fueled by hard-nosed science—based not just on climate change models, which by nature are to some degree speculative, but also on vast amounts of field observations. We have moved from predictive science to observational science.

Ideologues abound on both sides of the climate change debate, but one cannot deny or fail to be concerned by tide-gauge records of rising seas, temperature data showing warming oceans, observations of glacier retreat the world over, and maps showing the shrinking areal extent of permafrost in northern latitudes. Also, the "real problems" of poverty, terrorism, and broken financial systems that Dutton mentions will all be exacerbated by the realities of climate change.

We are already seeing climate change refugees (e.g., Pacific Islanders having to abandon their homeland because of sea level rise, and Alaska natives facing relocation because of erosion). In the next century rising seas will force many, including tens of millions of Bangladesh citizens, to seek higher ground. Even the economies of wealthy countries like the United States will be challenged, as billions of dollars will have to be spent to protect coastal cities. Weather patterns will be altered, causing drought in some places. Major rivers in Asia and elsewhere may disappear during the dry season, as global warming causes the high mountain glaciers to continue to melt.

The Petition Project

In a bulk mailing in 1998, the global warming denier Arthur Robinson, founder of the Oregon Institute of Science and Medicine, circulated what has come to be known as the "Oregon Petition." Attached to the petition was what appeared to be a publication of the National Academy of Sciences (NAS), using the same typeface and format as the NAS official proceedings and with a cover note signed by a former NAS president, Frederick Seitz, who is a physicist. This was not a peer-reviewed publication but a piece filled with misinformation, disguised as a legitimate publication in order to encourage scientists to sign the petition. The petition read as follows:

"We urge the United States government to reject the global warming agreement that was written in Kyoto, Japan, in December, 1997, and any other similar proposals. The proposed limits on greenhouse gases would harm the environment, hinder the advance of science and technology, and damage the health and welfare of mankind. There is no convincing scientific evidence that human release of carbon dioxide, methane, or other greenhouse gases is causing or will, in the foreseeable future, cause catastrophic heating of the Earth's atmosphere and disruption of the Earth's climate. Moreover, there is substantial scientific evidence that increases in atmospheric carbon dioxide produce many beneficial effects upon the natural plant and animal environments of the Earth."

The petition quickly picked up nineteen thousand signatures (the number was up to 31,486 by January 2010). The National Academy responded, noting that the "petition does not reflect the conclusions of expert reports of the Academy." It must be stressed that signing a petition in opposition to a concept is a statement of belief, and, as the retired climatologist R. G. Quayle pointed out in private correspondence, does not equate to peer-reviewed scientific assertions. Robinson has admitted to a lack of climate scientists on the petition. David McCandless and Helen Lawson Williams examined the background of the signatories and determined that 49% of those signing the petition are engineers. In contrast to the "no consensus" message of the Oregon Petition, Naomi Oreskes, in a review of the abstracts of 928 papers on global climate changes, found not a single one that did not explicitly or implicitly accept the human role. The Oregon Petition, floated as evidence that there is no scientific consensus regarding global warming, is nonsense.

THE FUTURE OF ICE ✷ 5

The World's Great Ice Sheets

The world has three great ice sheets: the Greenland Ice Sheet in the northern hemisphere and the East and West Antarctica ice sheets in the southern hemisphere. Each consists of a large central mass of ice ringed by a series of outward and seaward-flowing outlet glaciers. Glaciers are large masses of ice, which are frozen year-round and flow slowly and continuously down slope. The ice is created by the compaction and recrystallization of snow.

Ice sheets are the eight-hundred-pound gorillas of global climate change, because they will likely be the major drivers of sea level rise and changes in ocean currents, which in turn will have a large impact on global climates. No one really knows for certain what the future holds for these ice masses. For example, the amount of ice formed by future winter snowfall in the interior of the ice sheets remains an educated guess. Warming temperatures could lead to increased precipitation in the interiors. At present, however, the ice sheets are losing mass and melting at an ever-increasing rate. All indications are that they will continue to melt and that the melting rate will continue to accelerate.

Greenland is an island of 2.165 million square kilometers (836,000 square miles) with an ice cover of 1.753 million square kilometers (677,000 square miles), consisting of 2.868 million cubic kilometers (688,000 cubic miles) of ice up to three kilometers (two miles) thick. The ice that covers Greenland was probably first formed about two million years ago at the start of the Pleistocene epoch. In this epoch the Earth entered a period of cool atmospheric temperature, perhaps 6 degrees Celsius (11 degrees Fahrenheit) cooler on average than when the dinosaurs were kings of the animal kingdom. It is possible that the Greenland ice completely disappeared several times during interglacial times. Greenland is surrounded by the Atlantic Ocean to the southeast, the Greenland Sea to the east, the Arctic Ocean to the north, and Baffin Bay to the west. The weight of the ice has depressed the central part of Greenland into a deep basin a thousand feet deep. Thus if the ice were suddenly removed,

Greenland would be an island archipelago (until the land rebounded from the weight of the ice and recovered its elevation).

Antarctica is 98% covered with ice, which on average is about 1.6 kilometers (one mile) in thickness. The southern Indian, Atlantic, and Pacific oceans, known collectively as the Southern Ocean, surround it. The Antarctic continent is divided into West and East Antarctica, separated by the Transantarctic Mountains. East Antarctica is about the size of the continental United States and West Antarctica about the size of Texas.

The margins of the ice sheets are always melting as the marginal glaciers are pushed out to sea. It is clear now that the rate of ice loss at the outer edges of the ice sheets is greater than the rate of annual addition of ice by snowfall, so these ice sheets are experiencing a net loss of water volume.

The Greenland Ice Sheet has been melting for decades, but toward the end of the twentieth century the rate of melting increased. Probably until 2000 the West Antarctic Ice Sheet was essentially not losing mass at all and perhaps was even growing slightly (or so it was assumed by glaciologists). In the first decade of the twenty-first century this changed dramatically. Some researchers believe that the melting rate of the Antarctic Ice Sheet will eventually exceed Greenland's.

Atmospheric warming, at least at this point, does not affect the interior of the ice sheets, which are always well below freezing. Ice loss from ice sheet interiors is mostly by sublimation, a process by which dry snow or ice changes to water vapor.

Satellite observations are critical to monitoring the progress of the world's great ice sheets. Satellites make three types of measurements. The first is satellite altimetry, which simply measures the elevation of the ice surface. Satellite gravity measurements measure the mass of the ice cover. Changes in the mass are the all-important measure of loss or gain of ice. Satellite radar interferometry measures the velocity of moving ice and can also pinpoint the location of grounding sites for individual outlet glaciers. A grounding site is where the nose of a glacier is jammed up against an island or some sort of rise on the sea floor of the continental shelf.

Based on all of these types of satellite observations, the latest measurement of the rate of ice loss is 247 billion metric tons (273 billion short tons) of ice each year for Greenland and 120 billion metric tons (132 billion short tons) for West Antarctica. Moreover, in 2009 a report from J. L. Chen and associates at the University of Texas indicated that for the first time, the Antarctic ice sheet in the east experienced a net loss of ice, at a relatively small estimated annual rate of 52 billion metric tons (57 billion short tons). This ice loss is believed to have begun in 2006. This first report of a significant ice loss from the East Antarctic Ice Sheet is very likely a harbinger of a sea level rise larger than that indicated by current estimates. The combined melting ice contributions from the three ice sheets probably amount to between 30% and 40% of the total global sea level rise of around 3.2 mm (0.13 inches) per year. The relative importance of the contribution of ice sheets to sea level rise is expected to rise significantly in this century.

In 2010 a joint team of United States and Dutch researchers headed up by D. B. Dias concluded that previous measurements of ice loss from Antarctica and Greenland had overestimated the rate of ice loss, which was only half of current estimates. According to these authors, the rebound, or upward movement, of the land surface due to the loss of glacial ice weight was incorrectly accounted for in previous measurements. The authors admit that more field evidence is needed to substantiate their results.

If all the ice on Greenland melted tomorrow, sea levels would rise twenty feet or so. Estimates of the sea level rise potential of melt water from the West Antarctic Ice Sheet range from 3.7 to 5.2 meters (12 to 17 feet). If all three ice sheets melted completely, sea level would rise by 67 meters (220 feet). Obviously, if this were to happen coastal cities around the world would be destroyed. In any case these cities are potentially in serious trouble, even with very much smaller sea level rises.

How the Ice Sheets Work

The Greenland and Antarctic ice sheets are quite different. The Greenland Ice Sheet melts more or less in proportion to the warming of the atmosphere, but the melting of the West Antarctic Ice Sheet is more complex, and its future melt water contribution to the sea is not necessarily directly related to global warming. The rate of melting of all of the ice sheets is accelerating.

The rate of flow from the Greenland Ice Sheet is controlled first and foremost by the slope of the land and friction at the base of the ice mass. The release of water to the sea occurs as individual glaciers calve into ocean waters and as the surface and base of the glacier melt. During the summer the surface melt waters form shallow ponds and streams of water that plunge spectacularly through cylindrical moulins that lead through the glacier to its floor, lubricating the base of the glacier and speeding its seaward flow.

The complexity of the melting of Antarctica's ice sheets and the continent's status as the coldest place on Earth make it difficult to predict its contribution to the future of sea level rise. The same mechanisms at work on the Greenland glaciers also work here, but there are important additional factors. Many of the individual Antarctic glaciers that make up the margins of the ice sheet are grounded on the continental shelf, sometimes butting up against islands or other topographic irregularities. As a glacier thins from melting or retreats from calving into the sea, it may become detached from the shelf, causing a sudden acceleration of the ice flow into the sea.

Other Antarctic glaciers are buttressed up against ice shelves along the seaward

MOULIN, GREENLAND

Since the beginning of this century the melting of the Greenland Ice Sheet (which would raise sea level by twenty feet if the whole ice body melted) has been accelerating. Helping this process along are more or less cylindrical vertical shafts called moulins that allow meltwater to flow to the base of the glacier, lubricating the ice's path to the sea.

margin of the continent. An ice shelf is a floating sheet of ice two to four hundred meters thick; there are at least seventeen in Antarctica. The largest ones are the Filchner-Ronne ice shelf (430,000 km^2, or 166,000 square miles) and the Ross Ice Shelf (487,000 km^2, or 188,000 square miles, the size of France). Most large West Antarctic outlet glaciers terminate against these ice shelves. When ice shelves break up, the adjacent glaciers are "released" and their flow to the sea is accelerated. A recent, relatively minor example of such an event was the much-studied breakup of the Larsen B ice shelf (3,250 km^2, or 1,250 square miles), on the Antarctic Peninsula in February 2001.

A publication in 1968 by the late John Mercer, an Ohio State professor once described as a bold and eccentric glaciologist, may have been a very prophetic one, especially since global warming was little appreciated at that time. He argued that a major deglaciation of the West Antarctic Ice Sheet could happen within fifty years as the ice shelves began to break up, leading to the retreat of grounding lines, which in turn would lead to the disintegration of the ice sheets by calving into the seawater. He said that the breakup of the ice shelf on the Antarctic Peninsula would be the first step, and this first step has certainly happened, with the breakup of the Larsen and Wilkins ice shelves.

Two large outlet glaciers, the Pine Island Glacier and the Thwaites Glacier, empty into Pine Island Bay along the Amundsen Sea. These drain approximately 20% of West Antarctica and are the only two glaciers there that are not buttressed up against an ice shelf. The surface on which these are grounded is more or less a ridge, which means they are likely to become ungrounded as they shrink into the deeper water behind the ridge. Once they are ungrounded the calving is likely to increase, causing the glacier to retreat at a very high rate. This is why in 1981 Terry Hughes of the University of Maine called the Pine Island Glacier the "weak underbelly" of the ice sheet.

A third mode of Antarctic ice sheet disintegration, observed by Doug Martinson of Columbia University, is caused by the incursion of warm water onto the shelf of West Antarctica, which apparently only began in the last ten years or so. The water arrives through upwelling of deep water in response to winds that have changed direction because of global warming and possibly also because of atmospheric changes caused by the ozone hole. Warm water by Antarctic standards would not likely attract swimmers, since it is only a few degrees above freezing, but it is warm enough to increase melting at the margin of the ice sheet.

The Disappearing Alpine Glaciers

One of the most compelling and irrefutable lines of evidence of global warming is the global retreat of alpine glaciers. These glaciers, sometimes called mountain glaciers, are perennial masses of ice that occupy the valleys of high mountain ranges and slowly flow downhill toward lower elevations or to the sea, where ice can no longer survive. Almost all the world's alpine glaciers are currently shrinking, with the exception of glaciers at high elevations. The survival of an alpine glacier, just like the survival of the outlet glaciers of the ice sheets, depends on a so-called mass balance. The balance is the difference between the loss of ice in the summer and gain of ice through snowfall in the winter. By some estimates, if all alpine glaciers melt and disappear, sea level will rise perhaps one-half to two-thirds of a meter (one and a half to two feet).

As alpine glaciers begin to melt, the first thing that happens is thinning of the ice, followed by the more obvious retreat of the end of the ice lobe. The fastest shrinking by far occurs when the nose of the glacier calves in a body of water, although on land shrinking can be significant as well. The Khumba Glacier in the Himalayas has retreated more than 5 km (3.1 miles) since Edmund Hillary and Tenzing Norgay began their climb up the glacier on the way to Mount Everest in 1953. The Mendenhall Glacier near Juneau, Alaska, has retreated the same amount since 1760. But when a lake formed in front of the glacier and calving began, the ice retreated a full kilometer (0.6 mile) since 2000. Another impact of glacial retreat is illustrated by the dropping sea level at Juneau caused by the removal of the weight of the Mendenhall Glacier, allowing the land to rebound. The rate of movement of glaciers ranges from nil to several meters a day. The Jakobshavn Isbrae outlet glacier in Greenland is currently whizzing along toward the sea at 14 km (8.7 miles) a year.

The islands of the Canadian Arctic have a number of ice caps, all of which are shrinking and thinning. Ice caps are miniature ice sheets that form a central ice mass or dome from which smaller glaciers flow. Canadian examples are the Bylot ice cap on Bylot Island, the Barnes and Penny ice cap on Baffin Island, and the Devon ice cap on Devon Island. Other ice caps include the Vatnajökull of Iceland and the southern Patagonia ice field in Argentina and Chile.

A few floating ice shelves are associated with the Canadian alpine glaciers and ice caps. Although much smaller than their Antarctic counterparts, they have the same impact on glacier movement. When they break up, the glaciers formerly buttressed up against them immediately increase the rate at which they move toward the sea. Most of the shelves, including the Ward Hunt and Serson ice shelves on Ellesmere Island, are breaking up. The Markham ice shelf, also on Ellesmere Island and twenty square miles in area, broke away in its entirety almost overnight in 2008. What makes this event particularly interesting is that the ice shelf is known to be 4,500 years old, an indication that today's warming conditions in the Canadian Arctic may be quite extraordinary. The Ward Hunt ice shelf is estimated to be 3,000 years old.

One of the most worrisome aspects of the

GLACIAL CANYON, ALASKA

Alpine (mountain) glaciers could raise the sea level by one to two feet if they all melted. This glaciated valley has a characteristic U-shape in cross-section, in contrast to the V-shaped cross-section of valleys formed by rivers.

MOUNT MCKINLEY, ALASKA

Almost all the glaciers in Alaska are rapidly retreating. The few exceptions are some small glaciers at high elevations, such as those on the upper slopes of Mount McKinley, 6,166 meters (20,230 feet) tall. If all Alaska glaciers disappear, sea level will rise between 1 and 2 inches (25 and 51 millimeters).

shrinkage and loss of alpine glaciers is its impact on local water supplies and hydroelectric power. The most critical example of this is the retreating ice on the Tibetan plateau, which is the largest perennial ice mass outside Antarctica and Greenland and is sometimes called the third pole. Melting ice from more than 45,000 glaciers in the greater Himalayas furnishes crucial summer water to much of Asia. When the monsoons don't operate, this water is critical. Melting ice from the Tibetan plateau contributes to all the rivers between (and including) the Yellow River to the east and the Indus to the west, including the Ganges, Brahmaputra, Irrawaddy, Salween, Mekong, and Yangtze rivers.

By some estimates the water supply for close to a billion people in India and China is threatened by the melting third pole, but no regional cooperation or planning is apparent between the two giants, which instead are locked in border disputes.

In South America, La Paz (Bolivia), Ushuaia (Argentina), Quito (Ecuador), and Lima (Peru) are among the cities that depend heavily on glaciers for their water supply and also for hydroelectric power.

Melting glaciers very often have an international flavor. The borders between Italy and Switzerland and between Argentina and Chile are both partly on top of melting glaciers. The Siachen Glacier in the Karakoram Mountains extends across the boundary between Pakistan and India. This and other glaciers and their melt water resources are within the zones disputed by India and Pakistan. At present, water issues are not paramount, but the potential is there that melting glaciers will add a new element to the boundary disputes. Many glacial streams cross international boundaries, and often glaciers are responsible for recharging aquifers that may be international in extent. The Mekong River has a glacial source in China, and by the time it reaches the Mekong Delta, it has crossed China, Myanmar, Laos, Thailand, Cambodia, and Vietnam. Talk about a potential conflict over water.

Clearly the loss of alpine glaciers, which is accelerating, is a critical global issue, one that should be recognized and planned for immediately. Bhutan, a landlocked country squeezed between India and China at the eastern end of the Himalayas, was cited as the happiest country in Asia by *Business Week* in 2006. It has a government genuinely concerned with preserving the environment and the country's culture. It should not come as a surprise that Bhutan takes global change seriously and is making firm plans to move some small villages that are about to lose their water supply from soon-to-disappear Himalayan glaciers.

Not all agree, however. Richard Lindzen, the aforementioned MIT professor, believes that "since about 1970, many of the world's glaciers have stopped retreating and some are now advancing again." This simply is not true.

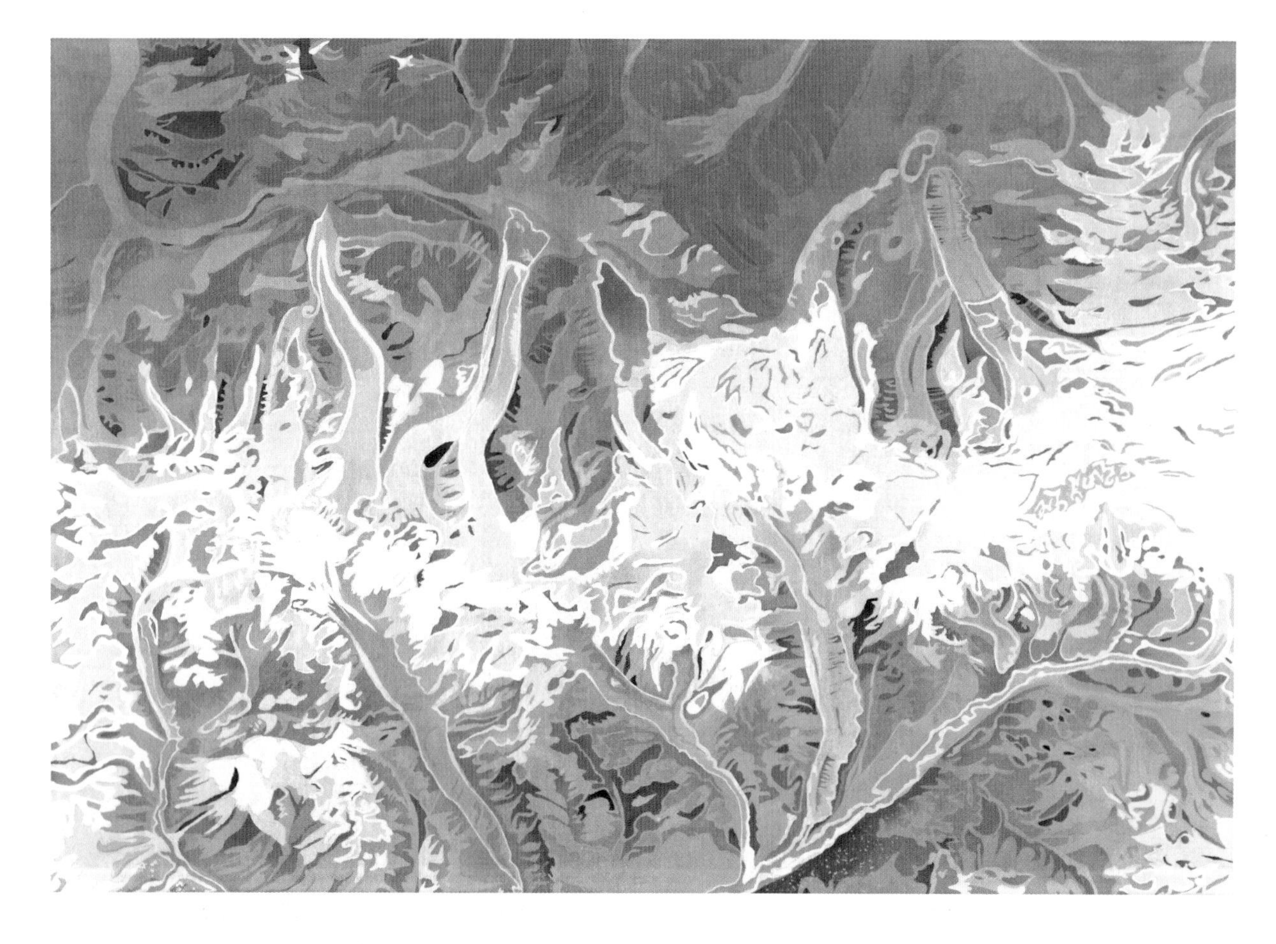

BHUTAN'S HIMALAYAS

The rapidly melting Himalayan glaciers furnish critical water supplies to most of the rivers in Asia. More than a billion people depend upon this glacial water, especially during periods of low rainfall in the region, but so far most regional planners seem to ignore the issue. One exception is Bhutan, where plans are afoot to move small villages because of a potential loss of water supply.

The Snows of Mount Kilimanjaro

Mountain glaciers are disappearing all over the world; perhaps 95% of them are shrinking, and those on Mount Kilimanjaro are no exception. Kilimanjaro is the highest mountain in Africa. At an elevation of 5,695 meters (19,340 feet), the mountain rises an imposing 4,572 meters (15,000 feet) above the plains that surround it. Like Mount Fuji and Mount St. Helens, Kilimanjaro is a strato-volcano consisting of alternating layers of lava and volcanic ash. Located in northern Tanzania, only 3 degrees of latitude south of the equator, it is one of the prime lines of evidence that global change is not just a phenomenon of the polar regions. The mountain's permanent ice cover must have been a jarring sight to the earliest Europeans traveling through the African tropics. Geologic evidence indicates that the ice cover has existed on the mountaintop without interruption since it first began to accumulate eleven thousand years ago. In recent decades the ice began to shrink dramatically, and by the beginning of the twenty-first century, between 80% and 90% of the mountain's ice had melted away. The remainder is projected to disappear within ten to forty years. The disappearance of the ice is widely believed to be caused by global warming.

The loss of the ice is a particular problem for the tourist industry of Tanzania. "The Snows of Kilimanjaro," one of Ernest Hemingway's most famous short stories, immortalized the mountain in the eyes of the western world and made the Kilimanjaro National Park one of Africa's prime tourist destinations.

Claude Allègre, geochemist and former French minister of education, penned an op-ed piece in 2006 for the French weekly *L'Express* in which he argued that the snows of Kilimanjaro are disappearing because of natural factors and that the cause of global warming is unknown. Allègre also wrote the book *L'imposture climatique* (The Climatic Deception, 2010) in which he downplayed the role of carbon dioxide in global warming, arguing instead that climate change had more to do with clouds and solar activity. His work aroused such controversy in France that the Academy of Sciences, of which he is a member, held a debate on climate change science, including the fate of the ice on Kilimanjaro. The Academy of Sciences issued a report in October 2010 in which it concluded that increased warming between 1975 and 2003 was mainly due to increased CO_2. The report stressed further that the increase in CO_2 and other greenhouse gases is unequivocally due to human activity.

The controversy over whether the loss of ice on Kilimanjaro is related to global warming illustrates a perception problem connected to global warming. It is much easier to observe and document climate changes on a continental scale than on a local scale. No one can reasonably deny that most mountain glaciers are shrinking and that this reality is a strong and irrefutable argument that the globe is warming. But

KILIMANJARO, AFRICA

The snows of Kilimanjaro, made famous in Hemingway's short story, are disappearing. This small, rare tropical glacier atop the mountain (5,695 meters, or 19,340 feet, tall) is rapidly becoming a highly visible victim of global warming.

when it comes to an individual mountain the case becomes less clear.

Nature is never simple, and undoubtedly other factors besides global warming are involved in the melting of the ice on Mount Kilimanjaro. These could include decreased precipitation; increased soot content of the ice, causing more absorption of heat from the sun; and local deforestation, which may have caused a reduction of moisture in the air. But the global loss of alpine ice and the persistence of the Kilimanjaro ice cover for thousands of years through a number of previous climate shifts (including droughts) argue strongly that global warming is the villain of Kilimanjaro.

Permafrost

Arctic lands are warming at five times the global average, and as in all things, there are advantages and disadvantages to such changes. The good thing about the warming of the far north is that it could lead to better farming conditions, improved shipping in the Arctic Ocean when the Northwest Passage opens, and easier transportation to markets and consumers. On a subtler level, musk oxen and caribou grazing, trampling, and defecating help to spread grasses, thereby attracting geese and adding to the productivity of lakes. Reindeer in Norway have benefited from less snow and longer growing seasons, which have reduced mortality and increased the birth rate.

The other side of the coin is that the warming is leading to permafrost melting, which in turn leads to beach erosion, the collapse of roads and buildings, the loss of pipelines and sewage ponds, and the release of large volumes of methane and CO_2. Many changes in flora and fauna are occurring. Some species are moving north and replacing indigenous species; for example, red foxes are displacing Arctic foxes. The encroachment of trees and shrubs onto the caribou tundra has reduced their food supplies. The calving season of caribou in low Arctic Greenland no longer corresponds to the time of maximum food availability, so fewer calves survive. Arctic plants start and grow earlier in spring and are past their energy-yielding prime when calving caribou cows need them. There is also reduced snow cover to provide insulation for hibernating species.

Permafrost is perennially frozen ground, both soil and rock, which underlies 20% to 25% of the Earth's land surface. It is particularly widespread in the Arctic. Permafrost does not always mean frozen water. It ranges from rock with frozen water between grains to soil with all pore spaces filled with ice. Much permafrost was formed thousands of years ago and is still maintained by current weather conditions. Permafrost extends to depths as great as five thousand feet in the Lena and Yana River basins in northern Siberia, but recently formed permafrost may be only ten feet or so thick. Although generally restricted to high latitudes, there is also an alpine form of permafrost, which exists at high elevations in mountain ranges (e.g., the Andes) at mid latitudes. Permafrost may be *continuous*, meaning that it is present everywhere with minor exceptions, or *discontinuous*, scattered about in patches. As a rule it stops groundwater flow and inhibits plant growth.

On top of the permafrost is a thin active layer a few feet thick, which thaws during the summer and allows the growth of a few species of plants. As permafrost melts and the active layer increases in thickness, plant diversity also increases. Plants flourishing in the new conditions may include bushes that can shade the surface and slightly reduce the rate of melting. The more diverse flora will increase the uptake of CO_2 from the atmosphere, but not at a rate that will counteract the newly released gases (as we noted earlier, primarily methane) from decaying organic matter. Animal populations are also affected by permafrost, because the hard ice restricts burrowing for dens. So reduction of permafrost may result in significant changes in the ecosystem.

Construction on permafrost requires special care to avoid melting, which causes the loss of strength and volume of the surface layer. Where melting has occurred, buildings have tilted, stands of trees have become "drunken forests," and roads and railroad tracks have collapsed. Of particular concern in China is the recently constructed Tibetan Plateau railroad, a project costing $4 billion and extending over 1,118 kilometers (695 miles), half of which is built on permafrost. Buildings in the Arctic are commonly built on pilings or on a three- to six-foot (one- to two-meter) layer of gravel. The Alaskan oil pipeline uses insulated pipes suspended above the ground between closely spaced towers to protect the pipeline from the fluctuations of the frozen ground. Melting permafrost also causes landslides in mountain regions and very extensive retreat of Arctic Ocean shorelines.

The most important global impact of shrinking permafrost is the release of methane and carbon dioxide (chapter 1). As permafrost melts, gas which was stored in bubbles in the ice is released, and even more gas is produced by the decay of long-frozen organic matter. In the Arctic, methane is also found as methane hydrates (frozen methane), which are held in place by the weight of the overlying impermeable ice in the permafrost zone. Although methane hydrates on land in the Arctic are much less voluminous than the same deposits on the sea floor, their release by melting permafrost could significantly add to the greenhouse gases released from permafrost.

Myths, Misinterpretations, and Misunderstandings of the Deniers

MYTH: *Scientists in need of funding invent reports of increased melting of permafrost.* Maria Leibman of the Earth Criosphere Institute in Russia denies that greenhouse gases affect the climate, and in a common ploy used by deniers, she suggests that the reports of increased thaw are from scientists in search of funding. Her views (showing that climate change denial is an international problem) would be laughable if the threat from the release of methane hydrates from permafrost were not so great. Of all the factors that could cause rapid, cataclysmic climate change, one of the most powerful could very well come from the release of methane deposits, causing what the British scientist E. G. Nesbit calls a "thermal runaway." This process begins with warming, which begets thawing in permafrost, which releases methane, which begets more warming, causing even more melting of permafrost and the release of still more methane—an accelerating, catastrophic sequence.

MYTH: *"The evidence suggests that the Greenland Ice Sheet is actually growing on average."* This is the view of Richard Lindzen of the Massachusetts Institute of Technology. As a member of the National Academy of Sciences, Lindzen is arguably the most distinguished of the global change skeptics. The field evidence, however, suggests that not only is the ice sheet melting but the melting zone along the edges of the ice sheet is expanding northward and toward the interior of the ice sheet.

MYTH: *Antarctica is gaining ice.* Well, yes and no. Land ice is decreasing in volume by around 163 billion metric tons (180 billion short tons) a year, which is responsible for a significant part of the sea level rise. Floating sea ice in the Antarctic winter (which has no bearing on sea level rise) is increasing in area, and this is often cited as an indication of cooling. In winter the sea ice surrounding the continent is about the size of Europe. But recent temperature measurements in the Southern Ocean indicate that the sea is actually warming at a relatively rapid rate (compared to mid-latitude oceans). According to a study by the British Antarctic Survey in 2009, the increased area of ice may be due to the ozone hole above Antarctica, which cools the stratosphere. The hole in turn has led to strengthened surface winds and larger storms, creating cold winds that emanate from West Antarctica. As the ozone hole is slowly repaired (perhaps by 2100) because of the successful global campaign to halt release of refrigerant gases, the production of winter sea ice will be reduced and additional warming of the Antarctic should occur.

MYTH: *The "West Antarctica Ice Sheet comprises only a small portion of Antarctica" and therefore is a minor contributor to sea level rise (according to James M. Taylor of the Heartland Institute).* Statements like Taylor's are apparently made in the hope that no one will check their veracity. The sea level rise potential in the West Antarctic Ice Sheet is a hardly minor sixteen feet.

MYTH: *The rate of melting of the ice sheets is slow and only presents a problem to humans on a millennial scale.* The observed rate of melting of the ice sheets, including the first recognition (2009) of a net loss of ice from the East Antarctic Ice Sheet, means that a sea level rise of one to one and a half meters (three to five feet) and possibly more will likely be achieved this century. A sixty-centimeter (two-foot) sea level rise will be a disaster for the world's developed sandy coastlines, as would a ninety-centimeter (three-foot) rise for many coastal cities. The most absurdly skeptical disregard of the importance of the world's ice sheets is that of Holly Fretwell of the Property and Environment Research Center. In a children's book, *The Sky's Not Falling*, she states that the melting of the world's ice sheets will not have much effect on global sea levels!

MYTH: *The glaciers of Kilimanjaro were receding when the planet was cooling in the mid-twentieth century, and therefore the loss of ice is not due to warming (according to Patrick J. Michaels of the Cato Institute).* Global temperature trends are not necessarily regional or local trends, which is one of the big problems facing scientists who try to predict local climate change. Thus even if Michaels is right and there was a mid-twentieth-century global cooling trend, it would not be an indication that there was cooling in central Africa. Distinguishing between global and local trends is critical in evaluating global change impacts.

MYTH: *Most of the 625 glaciers under observation by the UN's World Glacier Monitoring Service are growing. According to an article by the botanist David Bellamy in the New Scientist (2005), 555 of 625 observed glaciers have been growing since 1980.* The service's response: the evidence is unequivocal; most of the world's glaciers are melting. Apparently Bellamy's figures came from S. Fred Singer's website, but it's hard to comprehend how such a fundamental and now widely quoted error could have been made by a reputable scientist. Bellamy later recanted the numbers, but the damage was done and the numbers are still quoted.

GLOBAL CHANGE IN THE OCEANS ✳ 6

The Sea Level Rise

The level of the sea changes for many reasons and on many time scales. But melting land ice will be the major influence on global sea level changes on the time scale that should concern humans (chapter 5). The behavior of the world's ice sheets in the next four or five decades will determine whether sea level rise becomes the first great global catastrophe of climate change, flooding the coastal cities and displacing millions of rural dwellers in countries like Vietnam and Bangladesh.

Eighteen thousand years ago, a mere blip on the geologic time scale, the sea was 120 meters (400 feet) lower than it is today, thanks to the huge amount of water tied up in the massive glaciers and ice sheets grinding their way across the landscape from polar regions mostly in the Northern Hemisphere. Over the last two million years the sea level has risen and fallen in a major way at least seven times in rhythm with the advance and retreat of the glaciers. Currently geologists believe that we are in an interglacial stage, and we can expect the sea level to go down once more as the ice sheets re-form and advance toward the south. However, the return of the glaciers in full bloom may well have been delayed by warming of the atmosphere due to greenhouse gases.

After the last glacial maxima of eighteen to twenty thousand years ago, the sea level rose almost to its present level about four to five thousand years ago. From then on the sea generally rose very slowly to its present location. In different parts of the globe the rate of change differed, because the Earth is not a perfect sphere and also because the land surface is moving up and down in some locations. In the Southern Hemisphere sea level after the last glaciation actually rose two or three feet above the current level, and for the last two to three thousand years it has slowly dropped in places like Brazil, South Africa, and Australia. But with the sudden increase in greenhouse gases and warming of the oceans, sea level is rising almost everywhere today, and the rate of rise is accelerating. Currently the level of the sea is rising at an overall rate of 3.2 to 3.5 mm (1.26 to 1.38 inches) per year, as determined by both the last thirty years of tide gauges (a

CHARLESTON AIRBORNE FLOODED, SOUTH CAROLINA

Shown here is the area that will be inundated by a three-foot rise in sea level. More than inundation is involved, however. As sea level rises, tides and storm surges will penetrate further inland, storm drainage systems will fail, and groundwater will be polluted with salt water.

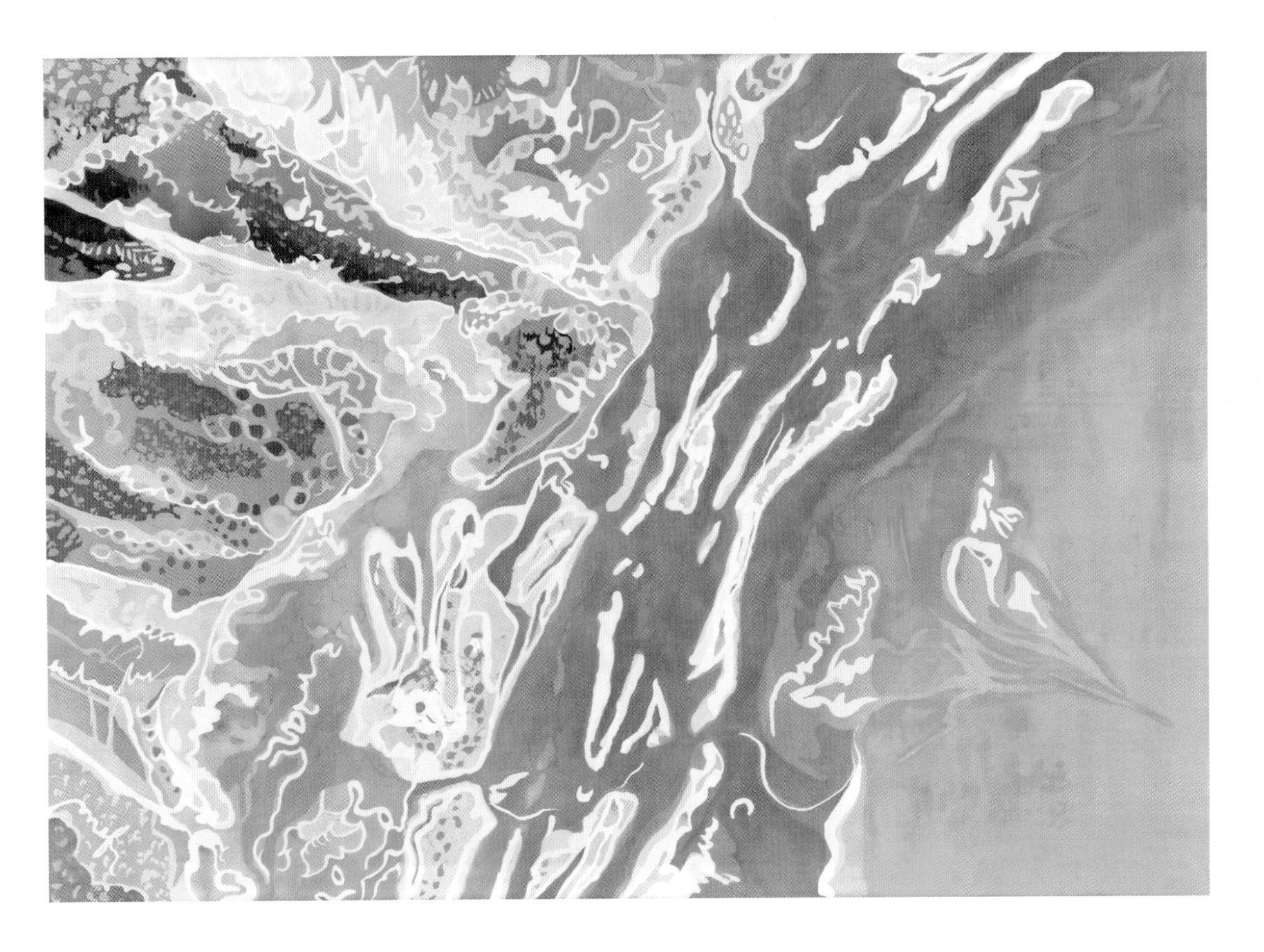

SPENCER GULF, AUSTRALIA

In the southern hemisphere after the end of the last ice age, sea level actually rose a meter or two above the present-day position and then gradually fell to its current position. Spencer Gulf on the Australia shoreline exhibits as many as six small barrier islands that were successively left behind, each at a lower elevation, as the sea level dropped over the last few thousand years.

total record of 150 years) and satellites (going back 18 years). This is more than 30 centimeters (1 foot) per century.

Sea level rise is actually a combination of the change in the volume of sea water (eustatic change) and the local up or down movement of the land (tectonic change). Eustatic change includes the increase in volume of the uppermost two thousand feet of the ocean water column because of warming water, and also the rise due to the introduction of "new" water from melting glaciers and ice sheets. Where the land is sinking, the sea level rise is enhanced. For instance, on parts of the Mississippi Delta where the sediment is compacting and the land is sinking because of the extraction of oil and gas, the rise is as much as 1.2 meters (4 feet) per century. Along the Pacific Coast of Colombia, the land at the coast is rapidly sinking, because of the forces actively building up the Andes mountain range. The coast sinks as the mountains rise, pushing the relative sea level rise rate up to as much as three meters (ten feet) per century. On the other hand, in the high latitudes the land may be rebounding from the recently removed weight of glacial ice. This is why the sea level is dropping in parts of Scandinavia, Nova Scotia, and Alaska.

During the twentieth century the primary cause of sea level rise was thermal expansion of the ocean water: an increase in the volume of seawater as it warms up. Roughly 80% of the warming by greenhouse gases is stored in the upper 760 meters (2,500 feet) of the ocean. The rest of the heat is in the atmosphere. Because of the downward momentum of the warm water through the ocean's water column, thermal expansion would continue for centuries even if global warming and surface ocean heating stopped tomorrow. The chemical oceanographer Philip Froelich likens the downward transfer of heat through the water column to a flywheel: like the flywheel in an engine, it provides momentum. As the cold deep water is heated, it expands. Simultaneously, the loss of this warm water that sank into the deep has a cooling effect on the shallow waters and causes the water there to contract. The expansion is greater than the contraction, so overall the ocean expands and the sea level continues to rise.

The next-biggest contributor to sea level rise over the twentieth century was the melting of mountain glaciers, followed by melting of the Greenland Ice Sheet. In the twenty-first century global change scientists believe that the Antarctic ice sheets will be the single biggest source of water causing the sea to rise, followed by the melting of the Greenland Ice Sheet, thermal expansion, and the melting of mountain glaciers.

The short of it is that we should expect a *minimum* of a one-meter (three-foot) rise in sea level by the year 2100, even in the unlikely event that we quickly reduce the amount of greenhouse gas emissions. The momentum furnished by the flywheel of sinking water will cause the system to roll on and on. Hal Wanless, a geologist at the University of Miami, believes that the most likely sea level rise will be 1.5 to 1.8 meters (5 to 6 feet) in the next hundred years. His number comes from an analysis of permafrost and sea ice changes, accelerating ice sheet melting, and increased thermal expansion. A rise of two meters (seven feet) is not out of the question, and prudent planners should assume the higher figure.

Ocean Acidification: One of the Evil Twins

Roger Revelle and Hans Suess, two geochemists at the Scripps Institution of Oceanography, envisioned the possibility back in 1956 that the ocean would turn more acidic because of excess CO_2. But this remained an academic curiosity until the last decade. Since then their suggestion has become reality in rapid fashion. In just the last five or six years this "new" greenhouse-related ocean phenomenon has come to the fore in the eyes of mainstream science. According to a seminal report by the Royal Society of London (2005), it has the potential for a more important impact on marine life than both global warming and overfishing. Recently the state of the science on ocean acidification was summarized in a series of papers in the December 2009 issue of *Oceanography*, and in 2010 the European Science Foundation presented a comprehensive policy briefing on the subject. Ocean acidification is sometimes called one of the evil twins of global ocean change, the other twin being sea level rise.

The process by which ocean waters are becoming more acidic is called ocean acidification, a term first used in the scientific literature in 2003. The average ocean water pH, or acidity, is now 8.1, compared to a pre-industrial pH of 8.2. On the steep logarithmic pH scale, that apparently small difference represents a 30% increase in acidity. Jelle Bijma of the Alfred Wegner Institute says that under a business-as-usual (BAU) scenario the surface waters of the oceans are likely to become 150% more acidic by 2100.

The delicate chemical balance which allows calcium carbonate ($CaCO_3$) to be extracted from seawater and used by all kinds of marine organisms to form shells and skeletons has been disrupted. A portion of the CO_2 that dissolves in seawater forms carbonic acid (H_2CO_3), which in large part controls the pH of the water. Carbonic acid is forming at a higher rate now because more CO_2 is being dissolved in seawater. Seas around the Arctic and the Antarctic are expected to experience acidification at a faster rate than other oceans, because cold water can take up more CO_2.

$CaCO_3$ is alkaline, or basic, the opposite of acidic. Because acidification makes the extraction of calcium carbonate from seawater by marine organisms more difficult, these animals will have a harder time making their shells, which will become thinner and more fragile. Certain critically important bacteria (such as the marine *Rosebacter clade*) that break down chemical compounds in the water as part of the ocean's chemical cycle will be reduced in number and effectiveness by even a small increase in acidity, which will greatly disturb marine ecosystems. The tiny organisms that make up plankton, some of the most important elements of the food chain, will be particularly affected. These include the coccolithophorids (plants), foraminifera (single-celled organisms), and pteropods (tiny snails also known as sea butterflies), all of which use $CaCO_3$ in their shells.

The pteropods are especially susceptible to acidification because their thin, fragile shells are made up of the relatively unstable $CaCO_3$ mineral aragonite. Pteropods are a base of the food chain, which extends from zooplankton to salmon to whales. In the Southern Ocean pteropods could disappear in this century, according to some studies. A number of commercial species including cod and salmon rely on pteropods in northern waters. A study in Alaskan waters estimates that a 10% decrease in pteropod abundance would lead to a 20% reduction in the weight of adult salmon that depend on plankton for food at one stage in their life cycle.

Coral reefs, which are also largely aragonite, will be damaged as acidification results in less robust, more vulnerable polyps. Disruption of the coral reef ecosystem, where fully one-third of all marine species exist, will have a particularly large impact on shallow water marine life over large areas of the ocean. Complicating the survival of the world's coral reefs is the simultaneous warming of ocean temperature and the rising sea level. Not helping matters is widespread human-caused pollution, plus events like the grounding in 2010 of a Chinese freighter running at full speed into the Great Barrier Reef of Australia and the oil spill in the same year in the Gulf of Mexico, which totaled around five million barrels.

There are two methods by which marine organisms build carbonate skeletons. The first involves scallops, oysters, some snails, and coral that exert little biological control over the precipitation of $CaCO_3$ in their skeletons. These creatures are dependent upon the saturation or super-saturation in seawater of the mineral they use, which means that they will be particularly susceptible to any acidification and reduction of carbonate in seawater. Very often the juveniles of these organisms are particularly sensitive to acidification. The second method of building skeletons involves animals having more biological control of calcification (as opposed to inorganic chemical control in the first method); hence the composition of seawater is less important to these organisms.

Of course, acidification involves lots of uncertainties, and there will be unexpected impacts. There is even some concern that acidification will proceed to the point where concrete structures in the sea such as piers and seawalls will be weakened by the dissolution of cement. Aquarium studies have shown that the giant Pacific Humboldt squid becomes sluggish and has difficulty breathing in acidic waters. A lab-

PTEROPOD (LIMACINA HELICINA)

Pteropods are tiny snails from open ocean waters that are critical parts of the food chain for a number of marine organisms including whales, cod, and salmon. These organisms, because of their mineralogy (aragonite) and very thin and sometimes transparent shells, are particularly susceptible to being dissolved by increasingly acidic ocean water.

oratory investigation by Justin Ries and associates at Woods Hole Oceanographic Institute observed shell growth in the laboratory under conditions of varied CO_2 content. Most shells were thinner under elevated CO_2 conditions, but seven of eighteen shelled organisms grew *thicker* shells. The researchers noted that soft clams and oysters showed reduced calcification, but at least in the laboratory some hard clams and lobsters and one species of coral didn't seem to care one way or the other about high CO_2.

A big ocean acidification warning came in 2010. Larval shellfish in hatcheries along the Oregon coast began to die, apparently because they were unable to form shells. Burk Hales, an oceanographer at Oregon State University, was able to demonstrate that the problem was the excessive acidity of waters pumped into the hatcheries from the open ocean. Here was proof that ocean acidification was at our doorstep and here to stay.

One of the main questions that must be answered is how carbonate-secreting organisms survived higher CO_2 concentrations in the atmosphere in the time of the dinosaurs. In the geologic past, as in the present, CO_2 existed in a natural equilibrium with a number of processes including the weathering of rocks, burial of organic matter in sediment, volcanic activity, and absorption in the ocean. Changes in any of those long-term parameters can affect the CO_2 concentration and hence ocean acidification. During the big jump in atmospheric temperatures about 55 million years ago known as the Paleocene/Eocene thermal maximum (PETM, discussed in chapter 1), a huge amount of carbon was released into the atmosphere, most likely from methane hydrates (frozen methane) from the deep sea floor. The resulting ocean acidification caused a huge mass extinction of deep-sea organisms (particularly foraminifera) as well as many land species. According to James Zachos and associates, the atmosphere and oceans took 100,000 years to recover from that change in climate. Its impact on shallow water organisms is, however, much less clear. Many species of shallow-water calcareous marine organisms survived the PETM, possibly because the weathering of rocks on land in a high-CO_2 atmosphere produced low-acidity river water that flowed into the broad, shallow seas of that time, counteracting the acidification caused by high CO_2 in the atmosphere.

The PETM events of millions of years ago (and others in the geologic past) played out very slowly, and as a consequence it is difficult to draw helpful lessons for today's very rapidly moving acidification crisis. It may be that the current rate of CO_2 addition to ocean waters will be a rare event in geologic history. There is a significant chance that this time acidification will affect the entire ocean water column, not just deep water as in the PETM event.

Melting Sea Ice and the Northwest Passage

The Arctic Ocean is the smallest and shallowest of the Earth's five major ocean basins. Floating sea ice covers most of it. Typically there are 15 million square kilometers (5.8 million square miles) of ice in the Arctic Ocean in the winter, which shrinks to 7 million square kilometers (2.7 million square miles) in the summer. Most of the icebergs and ice floes that break off of the sea ice remain enclosed in the Arctic basin, but a few escape, as the captain of the *Titanic* learned.

Since the beginning of satellite records in 1979, the Arctic sea ice cover, which follows a normal seasonal pattern of melting and refreezing, has been shrinking at a rate of about 11% per decade. At the same time, the average sea ice is getting younger. In 1988 31% of sea ice was five years old or older. That number dropped to 10% in 2009. The sea ice cover is also thinning, by 40% over the last few years. As measured by the aerial extent of the ice in the month of September (considered the time of the minimum "summer" sea ice extent), the years 2007 and 2008 were record minimums since 1997 (possibly because of particularly sparse cloud covers). The size of the summer sea ice cover varies, and in 2009 the ice cover was more extensive than in 2007 and 2008. Still, current predictions indicate that the Arctic ice cover will be entirely gone well before the end of the twenty-first century. Based on recent rates of sea ice loss, Muyin Wang and James Overland project a possible complete loss of summer sea ice in the Arctic by 2030.

As the ice cover is reduced, the albedo changes, and the ocean water warms all the faster. Albedo is the amount of solar radiation reflected by any surface, expressed as a percentage of the incoming radiation. Sea ice reflects back 30% to 40% of the incoming radiation it receives from the sun, while seawater reflects back only 2% to 10% of the radiation and absorbs the rest. Clouds may reflect back as much as 90%. The increased warming of the Arctic Ocean due to the albedo change leads to increased warming on adjacent lands, adding an additional threat to permafrost.

One of the exciting ramifications of the shrinking Arctic sea ice is the possible opening of the Northwest Passage, a shipping route along the northern tip of North America that would connect the Atlantic and Pacific oceans. Transits by submarine began with the voyage of the USS *Nautilus*, which traveled under the ice to the North Pole in 1958 as part of a complete transit of the Arctic Ocean. In 1959 the submarine USS *Skate* actually pushed through the ice at the pole. This passage was long sought by early explorers including Captain James Cook from the Atlantic side and Captain Vitus Bering from the Pacific side. A few ships including an ocean liner and a submarine (the USS *Seadragon*) have traversed the passage in recent years, and if Canadian sovereignty claims can be settled, the passage may become a summer reality in as little as a decade. Canada more or less views the Northwest Passage as its own Panama Canal, though no other nation agrees with this contention.

Meanwhile, for the first time, thanks to global warming, Canada has taken a strong interest in the remote far north, building small harbors and airstrips, and supporting mineral exploration, all in preparation for the disappearance of the sea ice. A few years back the Canadian government even established a tiny Inuit settlement in the far north to help territorial claims. The new settlers barely survived the experiment because of the lack of fish and game available for food at the ill-chosen site.

Sea Ice and the Southern Ocean

At the other polar extreme, large areas of sea ice are also found adjacent to Antarctica. Because the Antarctic is a large continent surrounded by vast, open ocean (unlike the enclosed Arctic Ocean), icebergs by the thousands drift far to the north every year. The sea ice extent during the Southern Hemisphere winter (September) is larger (17.9 million square kilometers, or 6.9 million square miles) than that of the Arctic, but the summer (February) ice cover is much smaller (2.8 million square kilometers, or 1.1 million square miles).

Unlike its shrinking Arctic counterpart, Antarctic winter floating sea ice has actually increased in area by 10% since 1980, a fact sometimes cited as an indication that the Earth is not warming but rather cooling. But the Southern Ocean is not cooling. Field measurements show that the ocean water is actually warming at a rapid rate. Hence it is likely that the increase in the area of sea ice is related to changing ocean currents, which are related in turn to changing wind patterns. Winds blowing offshore push floating ice away from the land. Thus as the Antarctic wind patterns change, the ice-free water patches relocate and penguin colonies migrate accordingly to get their food (krill) from the sea.

Ozone depletion in the stratosphere has very likely propagated downward and begun playing a role in altering atmospheric circulation and wind patterns. The winds that the ozone hole generates also increase salt spray in the atmosphere, which in turn creates bright clouds, partially shielding Antarctica from greenhouse gas warming. As the ozone hole closes and its effect on wind and ocean currents and cloud formation ceases, an increase in the rate of Antarctic warming can be expected.

In 2010 the columnist George Will created a storm of protest when he asserted that the extent of global sea ice at the time equaled that of 1979, implying that claims of melting sea ice were false. Technically he was right. However, he compared two months of ice-cover data (combining both Arctic and Antarctic sea ice) that were thirty years apart, a process that ignored a much bigger picture of gradual sea ice

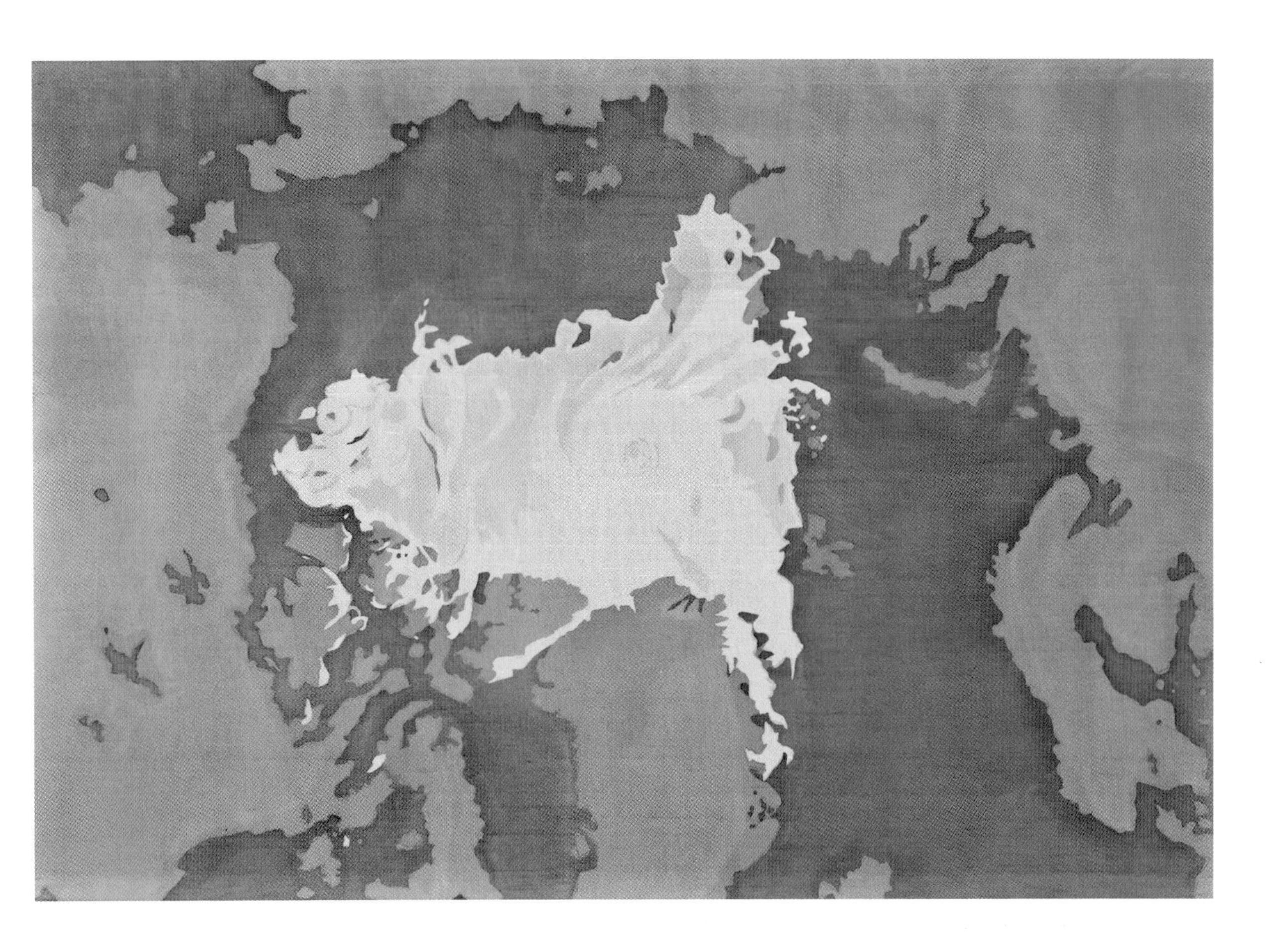

NORTHWEST PASSAGE

As a result of the gradual shrinking of the summer sea ice extent in the Arctic, the fabled Northwest Passage may soon connect the Atlantic and Pacific oceans, a great boon to international shipping. The decreased area of white ice will be replaced by darker seawater which will reduce reflection of solar radiation, further speeding up the warming of the Arctic.

reduction with a lot of annual variation. In addition, by combining Arctic and Antarctic sea ice numbers, Will was able to obscure the rapid decrease in area, age, and thickness of the Arctic sea ice. This cherry picking of data, or focusing on particular meaningless variations rather than the big picture, is a common tactic among climate skeptics.

Disappearing Arctic Sea Ice and the Animals

Loss of Arctic Ocean sea ice is causing problems for ivory gulls, the Pacific walrus, hooded seals, the narwhal whale, and polar bears, each for different reasons. The bears spend most of their lives on the ice and depend on it for capturing ringed seals as the seals surface at breathing holes. Five of nineteen polar bear subspecies are said to be in decline.

The plight of the polar bear, now classified as "threatened" by the United States government, has been given particularly wide publicity. It is an iconic animal, warm and fuzzy (from a distance). Lots of teddy bears are polar bears, but alongside a bear, walruses or hooded seals are rather unattractive, and few of them are found on department store shelves in the toy section.

The decision in 2008 to list the bears as threatened, based as it was on global warming projections that Arctic sea ice would soon disappear, was controversial. The state of Alaska opposed the designation because it would allegedly hinder future resource development. The state argued that the future of the sea ice was uncertain, that the models predicting ice trends were not accurate, and that the scientific literature on the bear was questionable. Hunting groups opposed the designation as well. One of the arguments of opponents was that the polar bears were believed to have evolved from grizzly bears perhaps 100,000 years ago. So why couldn't they just revert to their old food sources? The answer is that the change in ice is happening extremely fast in terms of the bears' food supply, and whether they can adapt quickly enough is questionable.

The polar bears are threatened because of their seasonal dependence on floating ice. The narwhal is threatened because it has specialized feeding habits and a very narrow range of habitat. The narwhal is a medium-sized tusked whale which weighs between 900 and 1,600 kilograms (2,000 and 3,500 pounds) and is capable of diving as deep as 1,370 meters (4,500 feet). In the winter the narwhal feeds on the life forms that spend the winter beneath the pack ice; no pack ice would mean no food.

The hooded seal spends much time away from the ice, but for its all-important birthing it requires pack ice. The pups, abandoned by their mother a few days after birth, reside for several weeks on the ice before venturing into the water on their own.

Less well known is the plight of the Pacific walrus, which still exists in large numbers. For the walrus the ice is important because it serves as a breeding place and a safe nursery for pups while adults dive to obtain food from the sea floor. When ice is absent the walrus may be forced to crowd along a shoreline: in 2009 as many as four thousand were found crushed on a Siberian beach. It was a disaster caused when the closely packed animals suddenly panicked and rushed toward the water. Normally the walrus are widely spread out on the ice pack, rather than on a narrow beach.

The ivory gull is an Arctic species that feeds on just about anything small that moves. The bird's food ranges from a variety of plankton to small fish to squid. The gull takes advantage of the rich assemblage of food found on ice margins and in ice cracks.

All these and many other animals require extensive sea ice for their survival. Will they be able to adjust to a new, ice-free world?

Myths, Misinterpretations, and Misunderstandings of the Deniers

MYTH: *Climatologists expect sea level to rise only slightly.* The petroleum geologist H. Leighton Steward says that most climatologists predict a sea level rise of seven or eight inches by 2100. The coastal engineer Robert Dean states that thirteen inches over the same time frame is the projection of hundreds of scientists. The geologist Lee Gerhard notes that we can expect a four-inch rise in this century. Bjørn Lomborg expects a thirty-centimeter (one-foot) rise. All four of these projections of sea level rise are extremely low relative to the IPCC projections, especially if one includes the likely contribution from melting ice sheets. In fact they are lower than the current sea level rise rate of well over thirty centimeters (one foot) per century, and there is not the slightest indication of a reversal of the sea level rise in coming decades. The problem is that the report issued by the IPCC in 2007 projected a very low sea level rise, because it disregarded the contribution of the ice sheets (chapter 3). Although the report noted the omission, stating that the effect of the ice sheets should be excluded because it couldn't be modeled, the deniers seem to have ignored this statement. Patrick Michaels and Robert Balling in their book *Climate of Extremes* absurdly characterize the IPCC's recognition of its omission of ice sheet melting as a small caveat. Small indeed! Meltwaters from the ice sheets are expected to be the most important source of sea level rise in this century. The implication that many researchers agree with the low projections is preposterous. At the farthest end of the spectrum is Nils Axel Mörner, a geologist at the University of Stockholm. He published a booklet refuting sea level rise with the title "Sea Level Rise: The Greatest Lie Ever Told." Mörner finds little evidence of global change and claims that some of the sea level changes recorded by tide gauges are more likely related to land movement, the local evaporation of seawater, and other factors. This absurd statement is widely quoted by deniers.

MYTH: *The IPCC estimate of sea level rise rate dropped between 2001 and 2007.* The 2001 report included an estimate of the contribution of the Greenland ice sheet. In the 2007 report the importance of the ice sheet contribution was recognized, but the projected sea level rise did not include either the Greenland or Antarctic ice sheet contributions. Thus the two reports are apples and oranges.

MYTH: *Ocean acidification is the next big hoax.* Alan Caruba, a freelance writer and public relations practitioner, has described global warming as the "Big Lie" and ocean acidification as the "next big hoax," and predicts that carbon dioxide will be a "boon in water as it is on land." Caruba's assertions have no basis in science whatsoever, and in the fashion of so many deniers he does not refute the scientific data (discussed above in this chapter) but simply skirts around them. The critical societal importance of ocean acidification has been recognized only recently, as discussed above, so we can expect larger and more organized denier guns to come to bear on this issue in coming years.

MYTH: *It's not the disappearance of sea ice that threatens the polar bears. Non-climate factors are causing the decline in polar bear population.* Dr. Willie Soon, a diehard denier, was the co-author of a non-peer-reviewed paper titled "Polar Bears of Western Hudson Bay and Climate Change: Are Warming Spring Air Temperatures the 'Ultimate' Survival Control Factor?" Soon noted in the acknowledgments that the paper was partially funded by "grants from the Charles G. Koch Charitable Foundation, American Petroleum Institute, and ExxonMobil Corporation." Why would the petroleum industry fund a review of research on polar bears? The answer is twofold. The companies have an interest in downplaying global warming and, in the case of polar bears, keeping a popular animal off the endangered species list so that the natural resources of the Arctic Sea region can continue to be exploited. It is important to note that the paper by Soon and associates is a review of research; it does not include original observations or data. There is nothing inherently wrong in review papers; they can be valuable additions to our understanding of some phenomena. But the deniers of global warming work exclusively with the data of others, rarely providing new observations.

In the article on the polar bear, Soon argued that there was no evidence of warming climate in western Hudson Bay and attributed the fall in polar bear population to other factors, especially hunting. In a response to this article, Ian Stirling (an expert on polar bears) and others rejected this argument and stressed that long-term trends in the population of polar bears in western Hudson Bay were consistent with the thesis that climate warming in western Hudson Bay was "the major factor" causing the sea ice to break up at progressively earlier dates each year. This shift has prompted polar bears to come ashore too fast, subjecting them to progressively poorer conditions for several months each year, impairing reproduction and the survival of young, sub-adult, and older (but not prime) adults. Stirling and associates noted that when the population declined the hunting quota for Inuit was no longer sustainable and that overharvesting likely accelerated the population drop.

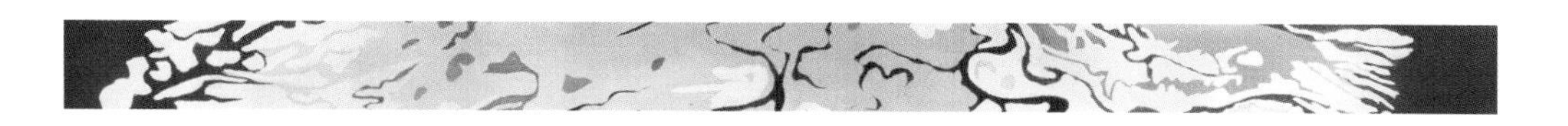

7 ✳ DISAPPEARING CIVILIZATIONS

The Problems with Coastal Living

What do Boston, Hong Kong, Shanghai, Lagos, Rotterdam, Alexandria (Egypt), Ho Chi Minh City, and Durban have in common? They are all major coastal cities in real danger of suffering major inundation from rising sea levels, and increasing damage from storm surge waves elevated by the higher sea level. These cities are but eight of thousands of low-lying communities, large and small, whose inundation by the sea may be the first catastrophe on a global scale caused by climate change. The human and economic costs of responding to the simultaneous flooding of all the world's low-lying coastal cities are almost unfathomable. Whether the solution is retreat, moving back buildings, abandonment, or construction of levees, dikes, and seawalls, it will divert much national treasure. And the high priority that will inevitably be given to protection of the major cities will take much national treasure away from the response to sea level rise in smaller communities and nations.

A significant portion of the world's human population has always lived next to or near ocean shorelines. In the last century this proportion has increased measurably. In fact, there would appear to be a global rush to the shore, and at the same time the shore is rushing toward the human invaders. Today in the United States, for example, 53% of the population lives on the 19% of the land area near the coast.

But there are flat coasts and steep coasts. The coastal zone of the western margin of North and South America, for example, is higher and steeper than its counterpart on the much flatter coastal zone on the eastern margin of the Americas. Thus the potential for damage from sea level rise is greatest on the east coasts of these continents. But all port facilities, critical components of the global economy, will be affected, whether on a steep coast (Long Beach Naval Shipyard) or flat coast (the Port of Miami).

Over the years a number of small East Coast communities have disappeared. Some have been lost to the waves, such as Edingsville, South Carolina, in 1893. Others have fallen into the sea, such as Broadwater, Virginia, in 1941. Still others were abandoned in the face of storm hazards: Diamond City, North Carolina, for example,

BOSTON II, MASSACHUSETTS

Greater Boston, with a population of around 4.5 million, is one of dozens of major urban areas around the world located at low elevations and highly vulnerable to the coming rise in sea level. Much of downtown Boston was created by filling in salt marshes and the nearby bay.

was abandoned and its buildings moved to safer sites on the mainland after three close calls with closely spaced hurricanes in the late 1890s.

Today lots of small nations and communities are in trouble. As beaches once locked in Arctic permafrost are released by the melting ice, native seaside villages are threatened by shoreline erosion as well as sea level rise and increased storm activity. The atoll nations in the Pacific and Indian oceans are already in the advanced planning stage of island and even nation abandonment.

Cities

MIAMI: NO HIGH GROUND

Globally the most threatened city of all (according to the UN's Organization for Economic Cooperation and Development), at least in terms of the value of property that will be flooded by a three-foot sea level rise, is Miami. With a population of a bit over 5.2 million, Miami is the fourth-largest city in the United States, behind New York, Los Angeles, and Chicago. It also is the nation's lowest city. Overall the average elevation is 1.8 meters (6 feet), but large areas are below 0.9 meters (3 feet) in elevation and the

EDINGSVILLE BEACH, SOUTH CAROLINA

In pre–Civil War South Carolina, Edingsville was a high-end resort community with sixty houses, two churches, and a tavern. The great Sea Island Hurricane of 1893 destroyed the houses, and subsequent erosion and island migration reduced the island to a narrow strip of sand less than a hundred feet wide. The old village, perhaps a harbinger, is now four hundred meters (one-quarter mile) offshore. Still, bits of brick, pottery and nails from the village often wash ashore in storms.

highest points are on the order of 6 meters (20 feet).

So what can the city do to prepare for the rising sea? Miami Beach, the barrier island fronting the city, will need to have seawalls on all sides. The beach will be long gone, no longer maintainable regardless of how much sand is pumped on it. The city proper will also need walls or dikes on all sides, including on the border with the Everglades on the western margin of the city.

But the geology underlying the city, as well as much of coastal southeast Florida, holds a surprise for the unwary engineer. Miami sits atop the Miami Limestone, a layer of sediment laid down during high sea levels over the last million years. This rock layer is up to fifteen meters (fifty feet) thick and is highly porous and permeable, meaning that water flows readily through cavities in the rock. According to the geologist Hal Wanless, evidence of the ease of flow of water through the limestone is found in the tidal fluctuations of the ocean that can be observed (on a very small scale) in ponds within the city.

The engineering ramification of this ease of flow is that mere walls or levees will not hold back the sea level rise. The level of the sea will simply be the same on both sides of the wall. Thus instead of walls, which are normally designed to protect the city from storms, dams will be needed, extending underground to depths that will depend on the thickness of the permeable rocks or sand underlying the city.

This is a problem that has only recently been recognized, and it is one in need of detailed engineering evaluation. It is a problem which probably exists along all 5,600 kilometers (3,500 miles) of the barrier island coast of the United States in the Gulf of Mexico, as well as along much of the Atlantic Coast. The major ramification of this is that holding the shoreline in place will be hugely more expensive than most current estimates of seawall costs. A massive wall, extending four and a half to six meters (fifteen to twenty feet) above the high tide line, will cost on the order of $33,000 to $82,000 per meter ($10,000 to $25,000 per foot), and the cost of a dam would be much more.

Miami has no high ground to which to move. The choices are to hold the line at great cost or to abandon the city at greater cost.

SHISHMAREF AND THE WARMING NORTH

One of the most immediate human impacts of the increased temperatures in the far north is the potential destruction of native seaside villages along all Arctic shorelines, including those in Siberia, Scandinavia, and northern Canada. In Alaska there are twelve such villages along or near the shoreline, in dire need of moving to higher, safer ground. Shishmaref, Alaska, is a typical village with typical problems related to the warming atmosphere. A small Inupiat Eskimo subsistence village just south of the Arctic Circle along the shores of the

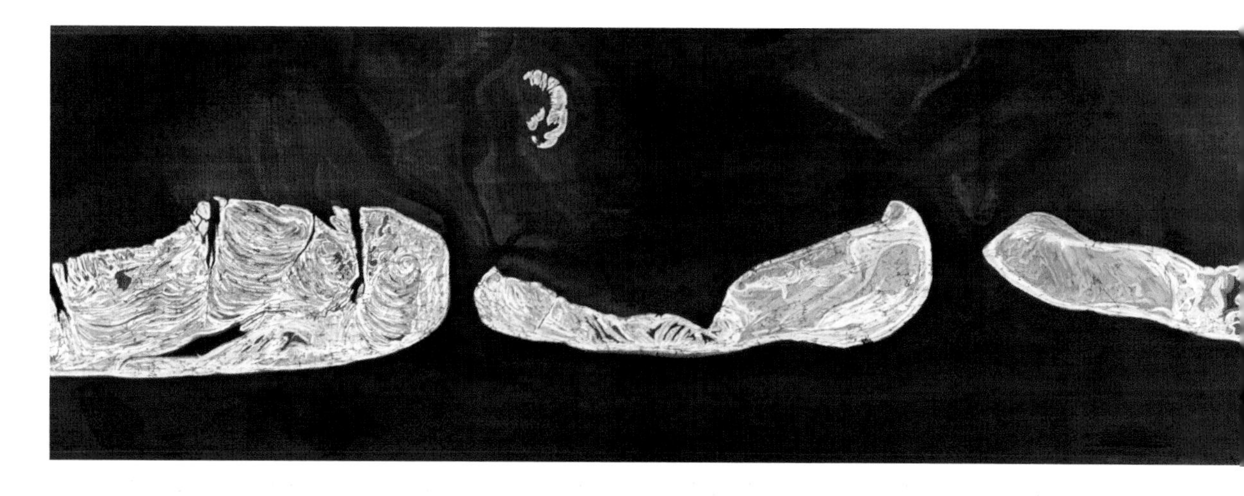

SHISHMAREF'S SHORES, ALASKA

The subsistence village of Shishmaref, with a population of 550, is threatened with destruction, principally because of longer ice-free periods of the adjacent ocean (allowing more storm waves to strike the shoreline) and the melting of permafrost in the beach sand. In the Arctic, moving an entire village, even a small one, is a very costly proposition. In this case relocation to the mainland is expected to cost $300,000 to $400,000 per villager.

Chuckchi Sea on Sarichef Island, an island six and a half kilometers (four miles) long, it has 550 inhabitants who live in small government-issue buildings.

In bygone days Shishmaref and the other villages became safely enclosed by a frozen ocean starting in September or October, so winter storms passed by harmlessly. Now the Arctic "summer" (the period of ice-free ocean waters) extends into November, so winter storms strike an island no longer encased in ice. In addition, the warmer climate and extended ice-free period have resulted in melting of the permafrost in the beach sand. Frozen beach sand acts like a natural sea wall, but when the grains are no longer held in permafrost ice, erosion accelerates. Adding to Shishmaref's woes is the rising level of the Arctic Ocean.

It has become clear that Shishmaref must be abandoned. But Inupiat Eskimo society is no longer as flexible or sustainable as it once was. There are several options for the village's response to sea level rise and warming climate. The villagers could move to the Alaska mainland, to another village, or to the "big cities" of Nome, Kotzebue, and Anchorage. These villagers may be skilled in hunting and fishing, but they have few of the skills useful in the modern city. The current plan is to move the village to the nearby mainland at a cost of $300,000 to $400,000 per inhabitant. It is a huge cost amplified by the price of providing water, sewer, heating, and electricity in the high north, but it could save one of North America's few remaining subsistence societies.

Atolls

Atolls are mid-ocean rings of coral rock that surround a lagoon. The origin of atolls was famously first described by Charles Darwin on the voyage of the *Beagle*. The islands began as a fringing reef completely ringing a volcano, and as the volcano slowly moved into deeper water (because of sea floor spreading), the reef grew upward and eventually extended above the now submerged mountaintop. These tiny islands became populated with Polynesians, the world's most impressive navigators, capable of traversing in outrigger canoes across thousands of miles of open ocean and finding new islands based on wave patterns. The islands became icons of song and romance and then of the battles of the Second World War. Now they have become the icons of the coming sea level rise catastrophe.

At 475 square kilometers (184 square miles), the largest atoll is Christmas Island, part of the Kiribati atoll nation. Most atolls have just a few square miles of land area and are very low in elevation. Usually the highest points are three to four and a half meters (ten to fifteen feet) above sea level along the open ocean shoreline, where storms pile up debris from the coral reef offshore. The living areas next to the lagoon are usually one to two meters (three to seven feet) above sea level.

Paul Kench, a New Zealand geologist, points out that atolls are dynamic but change at a slower pace than barrier islands (discussed below). Atolls widen as storm debris piles up and they lengthen as sand and gravel moves along beaches. Storms open new inlets between islands while often simultaneously closing other inlets.

The process responsible for driving people off the islands will most likely be the salinization of groundwater. Already in some communities in the Marshall Islands, crops are being grown in fifty-gallon oil drums to avoid salty soils. The Carteret Islands near Papua New Guinea have already been abandoned, and Tuvalu, a nation of five atolls with a population of just over twelve thousand, has made arrangements to relocate to New Zealand.

Perhaps the most famous atoll nation is the Maldives, a group of twenty-six atolls and more than eleven hundred islets in the Indian Ocean with a total population in excess of 300,000, making it by far the most populated atoll nation. The people of the Maldives have recognized the hazard of sea level rise in a big way. In 2009 the members of the Maldivian legislature donned scuba gear and held a session under water at a depth of nine meters (thirty feet) to draw attention to the sea level rise. The nation is currently contemplating the possibility of purchasing land in Sri Lanka or India and moving the entire nation to a new site.

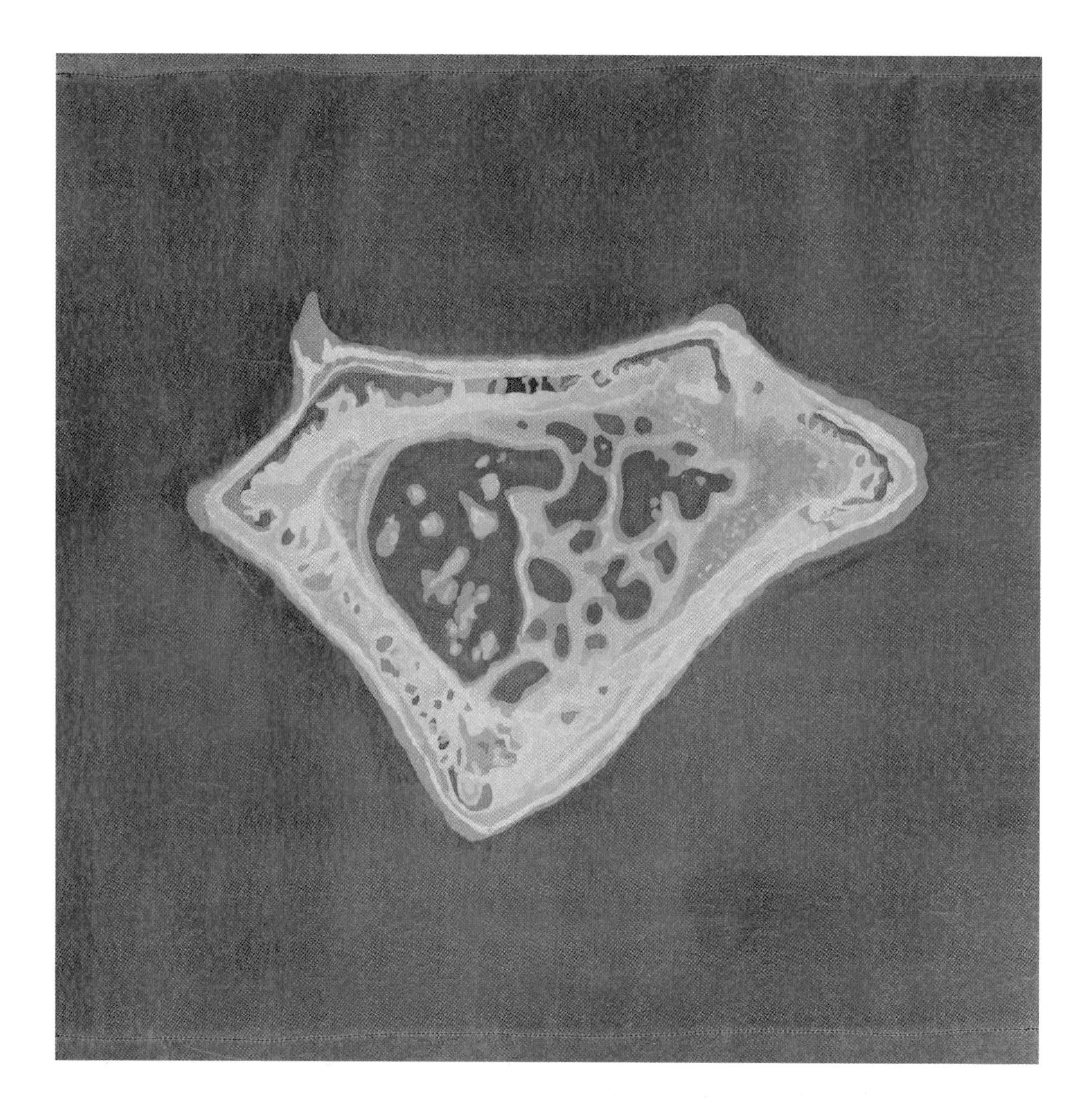

ATAFU ATOLL, TOKELAU, SOUTH PACIFIC

These tiny mid-ocean islands are the canaries in the mine of rising sea level. Before the sea inundates living areas (typically one to two meters, or three to seven feet, above present sea level) on the atolls, ground water contamination by salinization will likely drive the inhabitants away. Already some inhabitants are moving to higher ground.

Barrier Islands: Moving Ribbons of Sand

THE ECONOMIC FACTS OF LIFE

Barrier islands make up about 12% of the world's open ocean shorelines. They are a particularly vulnerable coastal landform, along with all kinds of sand spits and baymouth bars, and since they are popular sites for tourist development, their future in a world of warming and of rising sea level is of great concern. The greatest hazard facing these islands is an economic one. As the world's cities are increasingly threatened by sea level rise, it is an economic and political certainty that they will trump the more lightly developed barrier islands for funding. It also seems a certainty that the long-range fate of most of these islands must be abandonment. One exception may be the Persian Gulf city of Abu Dhabi, a city of one million inhabitants, the largest city located almost entirely on a barrier island.

THE OUTER BANKS OF NORTH CAROLINA

The Outer Banks of North Carolina on the East Coast of the United States is a chain of six barrier islands 210 kilometers (130 miles) long that extends from the Virginia border to Beaufort Inlet, North Carolina. The chain incorporates both Cape Hatteras and Cape Lookout and the two national seashores bearing their names.

As on the Portuguese Algarve barrier islands and the Dutch and German Frisian barrier islands, several villages coexist with the national seashores, causing some political discomfort. In 1972 the U.S. Park Service made the startling announcement that nature would be allowed to take its course on the seashores. It was a pioneering recognition of the ultimate futility of halting shoreline retreat in a time of rising sea level. It was also a recognition of the high environmental and economic cost of the coastal engineering approach to holding shorelines in place.

Barrier island migration, now recognized as a global phenomenon on coastal plain islands, occurs as the ocean side retreats and fans of storm overwash sand widen the back side of the islands. The ocean side retreats as the back side advances into the lagoon, and the whole island moves landward and upward, out of the way of a rising sea. Development on barrier islands of course halts the migration process. Of particular concern are high-rise buildings built adjacent to the eroding beach. In Florida there are virtually hundreds of miles of such high-rise-lined beaches, making impossible a rational response to sea level rise, such as moving back.

On most coastal plain islands, including the Outer Banks, the Frisian Islands of the Netherlands and Germany, and some of the Algarve Islands of Portugal, where the shoreline is not being held in place with engineering structures, shoreline erosion is occurring on both sides of the islands. The thinning rate of the Outer Banks of North

SOUTH OF OCRACOKE, NORTH CAROLINA

About 12% of all ocean shorelines are lined by barrier islands which, with the exception of the islands along cold Arctic shorelines, are in great demand for development. A one-meter rise in sea level will halt all development on these unconsolidated, low-elevation sand islands, unless they are ringed on all sides by massive sea walls.

Carolina is on the order of ten to fifteen feet per year. Clearly, thinning of these islands cannot have been a long-term phenomenon, or the islands would not exist today. Thinning is the initial response of a barrier island to sea level rise, and it will continue until the island is narrow enough to begin true migration, which can only occur when the island is thin enough (90 to 180 meters, or 100 to 200 yards) for overwash to frequently cross the entire island in storms. Typically islands in a full migration mode, such as those at Cape Romain (South Carolina), the Mississippi Delta, northern Yucatan, western Madagascar, and western Turkey, are less than one hundred meters wide. All these islands are responding in significant part to sea level rise.

Deltas: Where Rivers Meet the Sea

Deltas are the bodies of sediment that settle out when rivers flow into a standing body of water like a lake or ocean. Viewed from the air, the meandering and ever-splitting river distributaries and the intervening marshes and sandbars on deltas form beautiful patterns such as those of the Yukon Delta in Alaska and the Selenga River Delta in Lake Baikal, Siberia. It is because of an abundance of water and fertile land that many of the world's great civilizations originated on deltas, including those on the Nile, Rhine, Indus, Pearl, Ganges-Brahmaputra, and Tigris-Euphrates rivers. Today many deltas remain as population centers. The two most populous deltas are the Ganges-Brahmaputra (111 million inhabitants) and the Mekong (47 million).

Deltas began to form when the rising sea after the last ice age reached close to the present sea level six thousand years ago. Over the last two thousand years the growth of deltas has accelerated, because of the vast supply of sediment eroding from farmers' plowed fields. More recently, however, the land on most deltas has been sinking. Localized maximum rates of sea level rise on deltas range from four feet per century in parts of the Mississippi Delta, to 2.4 meters (8 feet) per century on the Bengal delta, to 3 meters (10 feet) per century on some small deltas on the Pacific Coast of Colombia, to as much as 10 meters (33 feet) per century on the river promontories of the Nile Delta. The sinking of land causing such high rates of sea level rise is due to the natural compaction of muds, often exacerbated by oil and water extraction, and the construction of dams upstream, which reduce sediment supply. The construction of canals on deltas removes sediment-trapping marshes and mangrove swamps that help to add land and elevation on deltas. Sinking (and rising) may also be related to tectonic forces within the earth, particularly near active mountain ranges. An example would be the river deltas on the Pacific shores of Colombia and northern Ecuador, which sink simultaneously as the nearby Andes Mountain Range is pushed upward.

As sea level rises, the potential for displacement of people is larger on deltas than any other place on Earth. Deltas are low in elevation and, as a consequence, are susceptible to storms, tsunamis, and particularly to sea level rise. Cyclone Nargis, which killed at least 150,000 people on the Irawaddy Delta in Myanmar in 2008, is only the latest delta catastrophe. The rising sea will also be particularly hard on agriculture in deltas. A study by X. Chen of the impacts on the Yangtze Delta predicts that rising seas will prolong the waterlogging of fields as groundwater levels are raised. During dry periods saltwater intrusion will occur on an ever-larger scale above groundwater levels.

Fifteen million people live at elevations within a meter of sea level on the Ganges-Brahmaputra Delta (the world's largest delta), most of whom will become refugees within the next fifty years. In wealthier societies it is likely that billions of dollars will be spent to fight the rising sea on deltas. Holland, which occupies the Rhine-Meuse Delta, will most likely succeed through extensive, massive, and costly engineering efforts. Globally, all delta communities (with the possible exception of Holland) will be abandoned within this century. New Orleans is a goner.

The three global delta hot spots as measured by the number of people that will be forced to leave their homes by a one-meter (three-foot) sea level rise are the Mekong, Ganges, and Nile deltas. Rick Tutwiler of the American University in Cairo describes the Nile Delta as a "kind of Bangladesh story"—on top of all the population, water, and pollution problems "there is the rising sea—a perfect storm." Based on a World Bank study of eighty-four coastal areas, the rise in Vietnam will displace a third of the population in the Mekong Delta, where half the country's rice is grown. A full third of Ho Chi Minh City would be inundated by a one-meter (three-foot) sea level rise. Vietnam's Red River Delta in the north will also suffer population displacement and loss of rice production.

Unfortunately in Vietnam the situation just got more complicated. A major dam has just been constructed on the Mekong River in China. As has happened on the Mississippi, the Nile, and the Niger River Deltas, the loss of sediment, which will be trapped behind the dam, will cause shoreline erosion and delta sinking. Adding sea level rise into the mix only intensifies these problems.

Myths, Misinterpretations, and Misunderstandings of the Deniers

MYTH: *Coastal engineering will save the day and hold shorelines in place and prevent inundation!* In a coastal management newsletter in Florida in 2010, readers were assured that engineers could solve the sea level rise problem. Bjørn Lomborg in some of his writing implies that the sea level rise problem is solvable, presumably by engineering. Although engineering could "save

BANGLADESH

Around fourteen million people will have to move to escape a one-meter rise in sea level here in the Ganges-Brahmaputra Delta. These refugees will have to be resettled in what is already one of the world's most densely populated countries. National boundaries prevent the settling of refugees into more sparsely populated regions nearby, a problem that can be expected to lead to local wars.

MOUTHS OF THE MEKONG, VIETNAM

Inundation of the Mekong Delta as well as the Red River Delta to the north will displace a larger proportion of the population in Vietnam than in any other country. New dams constructed on the river by China complicate the situation there in the same fashion as did the Aswan High Dam in Egypt. Less sediment and fewer nutrients will be delivered to the coast, shoreline erosion will increase, the fishing industry will collapse, and refugees will crowd into nearby cities, in this case Ho Chi Minh City, one-third of which would also be inundated.

the day" for a century or two, in practice it won't. The cost will be too high. Beach nourishment sand is not likely to stay on a developed beach which has had a one-meter (three-foot) sea level rise. Seawalls, extending to depths sufficient to prevent saltwater intrusion, will be too costly except near major cities.

MYTH: *Overall global change is a good thing. For example, carbon dioxide might help plant growth, atmospheric warming will make some portions of the world more habitable, the Arctic Ocean will become navigable, etc., etc.* Perhaps all these assertions are true, but the positive aspects of global change cannot be viewed in a vacuum. Sea level is definitely rising and coastal cities are threatened. Will the alleged positive aspects of increased atmospheric CO_2 and warmer temperatures offset the sea level rise catastrophe facing coastal cities and the acidification crisis facing marine organisms?

8 ✳ GLOBAL CHANGE IN THE BIOSPHERE

Plants and Animals of the Nearshore Zone

Two important groups of organisms live in the zone between the high and low tides and in the world's shoal waters just offshore. These are the salt marshes and mangroves. A third organic community, the coral reefs, lives in shallow water below the low tide line. All have several things in common. They are all important and widespread habitats for large numbers of species of marine organisms. All three are quite capable of moving landward or seaward as the sea level rises and falls, and they have historically migrated to the north or south as nearshore waters cooled or warmed during past climate changes. Unfortunately the three communities have another thing in common: their vulnerability to the activities of humans. All are threatened by global change and even more by the response of humans to global change. Like so many biological systems, left to their own responses they would probably survive quite well.

Generally mangrove forests and salt marshes are found on shorelines that are protected from direct wave attack. They often line the shores of estuaries, sounds, and lagoons or, as in the Great Barrier Reef, in the sheltered waters behind coral reefs. Coral reefs are found on open ocean shorelines and thrive in the open ocean wave environment, where their location is often marked by a telltale offshore white line of breaking waves. Coral reefs and mangroves are found in warm waters, while salt marshes flourish from the temperate zone all the way north into the Arctic.

Threatened by Acidification, Warming, and People: The Future of Coral Reefs

Tropical coral reefs exist in a broad band circling the globe, between latitudes 30 degrees north and 30 degrees south. They are found in clear, well-lit, warm, shallow water. Globally an astounding three million species of marine organisms live in, on, or very near the world's coral reefs, perhaps one-third of all such critters. The greatest of all is the Great Barrier Reef of Australia, 2,000 kilometers (1,240 miles) long. In the absence of humans, coral reefs should

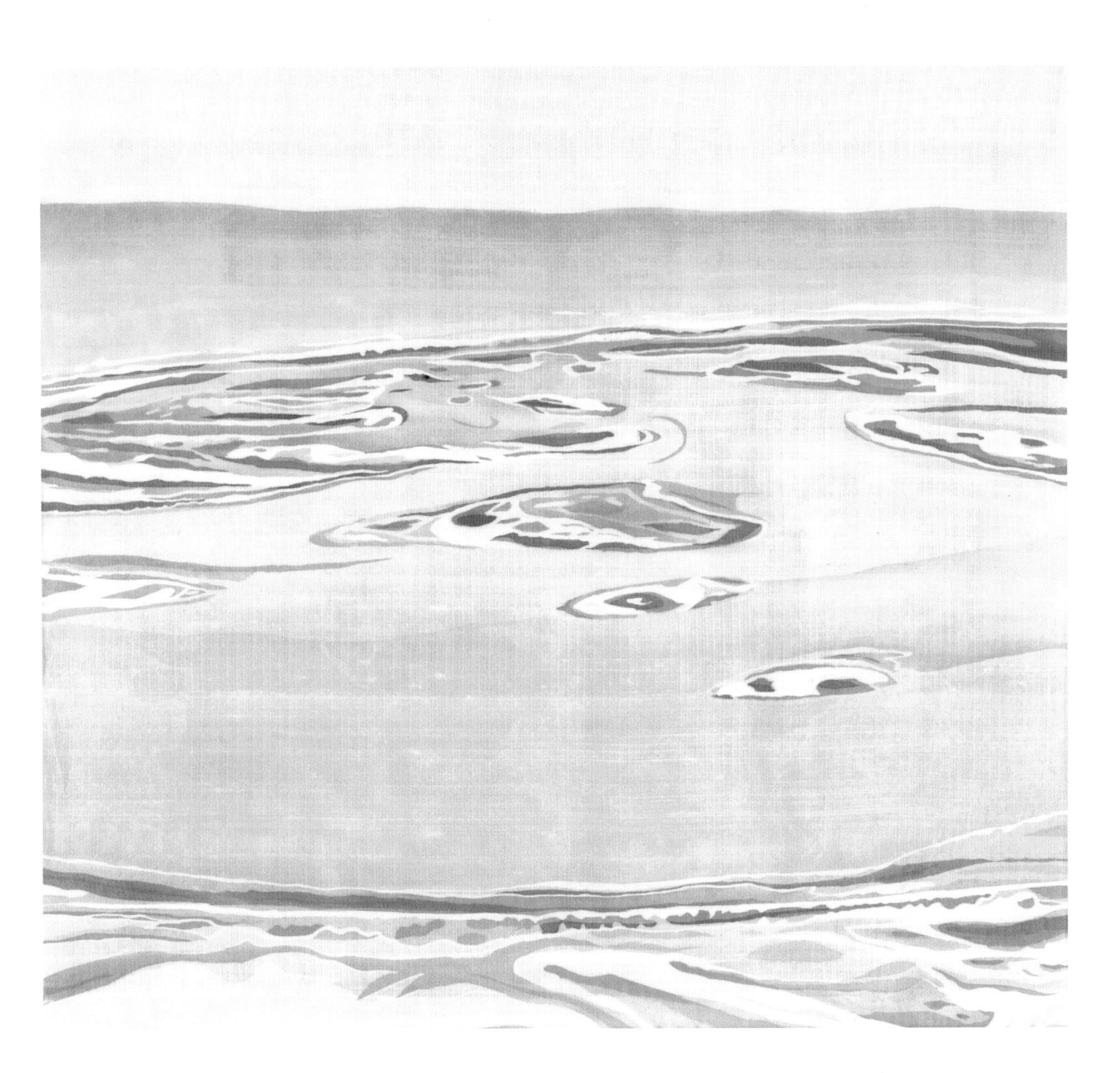

LAGUNA MADRE, MEXICO

Within Laguna Madre, Mexico, are more than six hundred small islands, most of which are surrounded by small marshes. As sea level rises, these islands, along with the marshes, will disappear, but this presents no problem to local people because the islands currently are uninhabited.

GREAT BARRIER REEF II, AUSTRALIA

Coral reefs everywhere are under attack by changing ocean conditions such as warming and acidification. Greatly adding to the problems of reefs are the impacts of humans who pollute the oceans, mine the reefs, and bury them in sand that can wash over them from nearby artificial beaches.

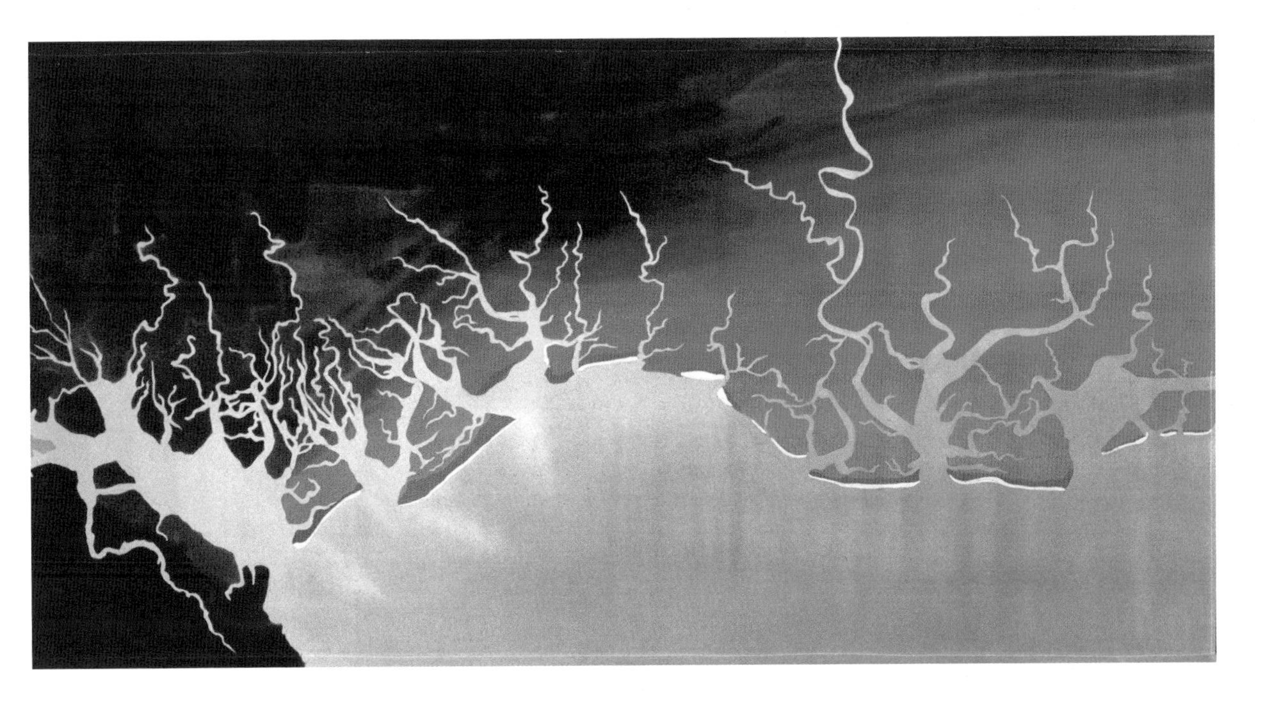

SINKING COLOMBIAN SHORES

Mangroves are the warm-water equivalent of salt marshes. Along the Pacific Coast of Colombia they form a continuous swamp-forest from the mainland to the oceanfront of barrier islands. Before the 1990s these barrier islands were not recognized as such, because the dense forest concealed the lagoon behind the islands.

be able to successfully keep up with the expected sea level rise by migration into shallower water or upward growth.

Yet the likelihood of reef survival is greatly reduced by the stress caused by humans. The array of hazards, both manmade and natural, threatening coral reefs is seemingly endless. The increased carbon in the atmosphere is causing ocean acidification, which is probably weakening coral skeletons and making calcification difficult for juvenile corals. If temperatures rise too quickly, the result can be bleaching caused by the loss of oxygen-producing zooxanthellae algae, and death for many corals and their associated organisms. Humans drive ships that crash into reefs, dredge channels through them, drag anchors over them, spill oil on them, cover them with beach sand, mine them for building blocks, pollute the waters, and collect coral heads for mantelpieces. The loss of the offshore breakwaters that reefs form will lead to greatly increased rates of mainland shoreline retreat.

The loss of reefs will represent a loss of an important carbon sink, leading to increasing carbon concentration of the atmosphere. The loss of the reef fauna and flora will be a loss to the fishing economy of many local communities and the economy of numerous tourist villages. Most important will be the loss of a huge number of reef-dependent species of marine organisms.

Mangroves: Disappearing Forests

Mangroves occupy a band circling the globe roughly between 25 degrees north and 25 degrees south and are the warm-water equivalent of salt marshes. There are at least thirty-five species in this group of plants that can tolerate salt water to varying degrees. They are referred to as mangrove forests or mangrove swamps. Species numbers are quite unevenly distributed. For example, there are only three species of mangroves in Florida Bay and perhaps nineteen species in Australia. Along shorelines, individual species are typically arranged in shore parallel bands, determined by the species' tolerance for tidal inundation, salinity, and waves.

Mangroves range in size from bushes to trees as high as a hundred feet, with tangled, almost impenetrable root systems. So far humans have done much more damage to mangroves than sea level rise and global warming have done. Mangrove wood is used as lumber and firewood, and large areas of mangroves have been cleared to allow access to and a view of the sea. Mangroves are considered ugly and are inevitably cut down and replaced by palm trees, as in south Florida during the 1920s. The biggest global threat of all is the clearing of mangrove forests to make way for shrimp farms.

Like salt marshes, mangroves perform important functions as a source of nutrients to local waters and as a shelter (within the root system) for a variety of organisms. At the top of the mangrove forest food chain are jaguars in Colombia and tigers in Bangladesh. Mangroves have proved effective in reducing damage from storms and tsunamis. Mangroves prevented wave attacks on buildings in Homestead, Florida, from Hurricane Andrew in 1995, and the Rangong area of Thailand received relatively little damage from the Asian tsunami in 2004 because mangrove forests remained intact. As sea levels rise, mangroves must move inland with the retreating shoreline, and as climates warm they must move to the north above the equator and to the south below the equator. Development will usually prevent this. On the other hand, most mangroves have very efficient methods of long-distance seed dispersal, which is certainly a strong factor in favor of survival. But the endurance of continuous forests along the world's tropical shorelines seems a faint hope, except for remote, undeveloped shorelines.

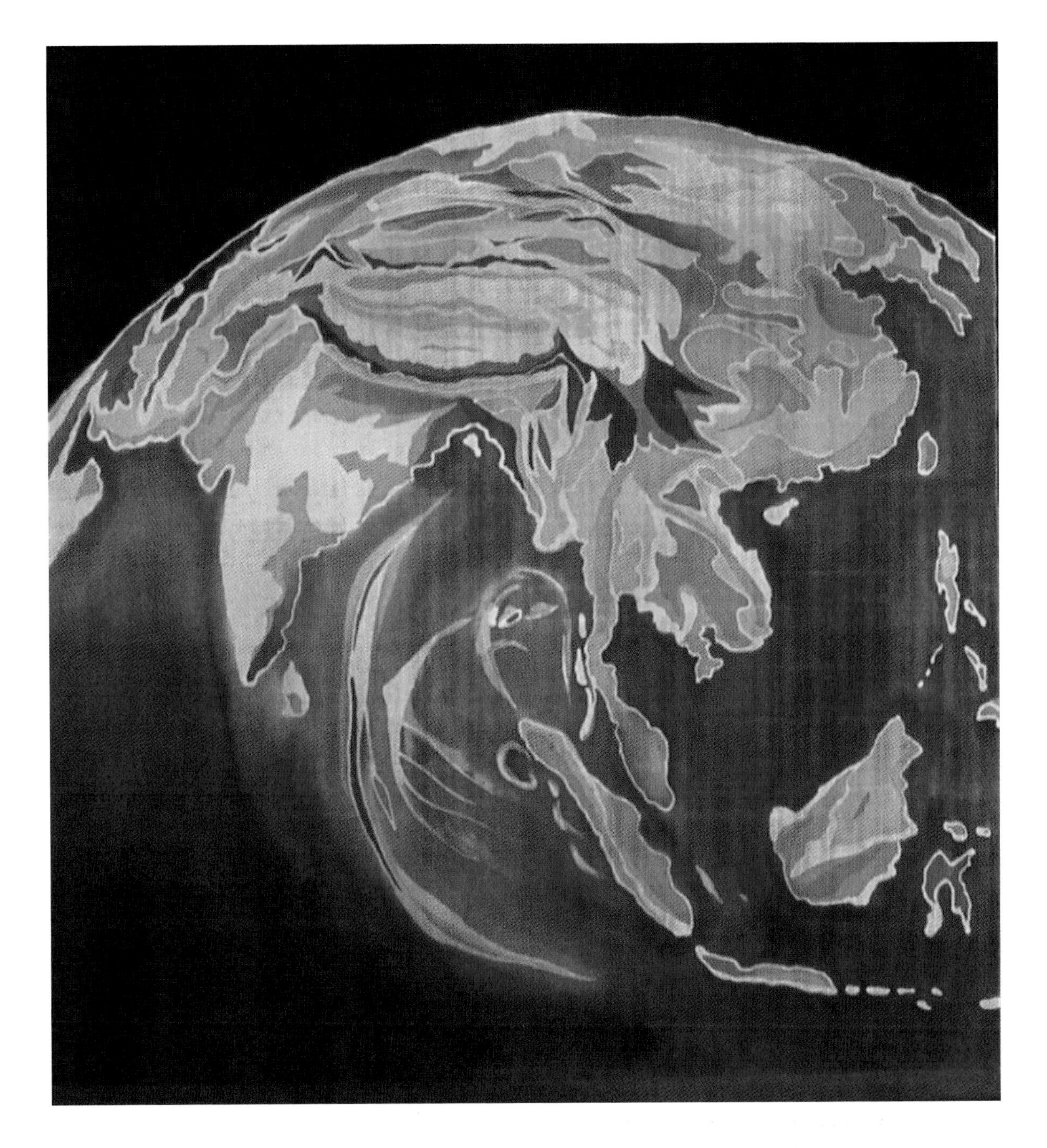

TSUNAMI

The tsunami in the Indian Ocean that killed 238,000 people in 2004 proved the value of preserving mangrove forests along shorelines. In the few places where these forests existed along affected shorelines, the tsunami wave was much reduced in power and lives were saved.

Salt Marsh: Nowhere to Move

Salt marshes, sometimes known as tidal marshes, live on flat areas, covered with a continuous carpet of salt-tolerant vegetation, next to shorelines between the high and low tide lines. Salt marshes range in width from a meter or two to several miles. The size of the marsh may depend on the tidal amplitude: the higher the amplitude, the wider the marsh. In river deltas the areal extent of the marsh depends on the size of the deltaic platform built by the river.

Salt marshes are characterized by extraordinarily low plant diversity, because few plants can tolerate saltwater. They are highly productive environments, creating through decomposition nutrients that feed a long chain of organisms. "Low marshes," which are inundated by almost every tidal cycle, generally have a single plant species, which on the shorelines of the eastern United States is the smooth cordgrass *Spartina alterniflora*. Plant diversity increases slightly in "high marshes" that are inundated only occasionally by saltwater. Beyond the high marsh in long estuaries or on very low-lying land are freshwater marshes, some of them huge, like the Florida Everglades.

Humans have created many problems for salt marshes. U.S. East Coast salt marsh grass has become an invasive species in Washington State and is taking over the mudflat environment, edging out local plants and animals. Ditto for the European salt marsh grass in New Zealand. Globally many square miles of salt marsh have been eradicated to make way for urban development, recreation (e.g., marinas), farming, aquaculture, salt making, and industry (especially in Taiwan). Boston, San Francisco, Tokyo, and Rotterdam have all expanded over salt marshes. And of course storm water runoff and sewage create local but widespread problems for the health of salt marshes.

Adding to the woes of the world's salt marshes is the sea level rise. Marshes can handle the rise with ease: they simply move back. Unfortunately the land next to salt marshes is highly treasured by humans, and inevitably such land is "protected" from erosion, which means that the shoreline is not allowed to retreat. Thus as the water level rises, the marsh narrows and will eventually disappear. The impact of the rising sea level is manifested along many of the world's estuarine shorelines by the line of dead trees whose roots have been flooded by either fresh or saltwater as the water table moves up apace with the sea.

Land Plants and Animals

Anecdotes of changes in the life cycles of plants and animals, apparently related to global change, are legion. Perhaps the most famous of these changes is the displacement of the polar bear (chapter 4), which in some areas of the Arctic is losing the sea ice essential for its seal hunting. Polar

YUKON DELTA, ALASKA

The Yukon Delta in Alaska consists of large areas of saltwater and freshwater marshes separating various river branches (distributaries). Salt marshes have proved to be important habitats for a variety of marine organisms, especially at their juvenile stages, which is why their preservation is important.

FLYING THE EVERGLADES, FLORIDA

The Florida Everglades are the largest and most widely recognized marsh and mangrove system in America. Besides being threatened by sea level rise, the Everglades are under pressure from agricultural activities, swamp draining, road construction, and the threat of invasive species such as the Burmese python.

bears have a problem because of habitat destruction. Another response of gigantic proportions is the saga of the spruce bark beetle (*Dendroctonus rufipennis*), which has killed spruce trees in a three-million-acre zone of south central Alaska. The problem is warmer winters, which have not allowed the normal winter die-off of beetles, and warmer summers, which have promoted faster reproduction of the beetles.

Plants

One big problem is how to separate the effects of the many hazards facing plants, such as invasive plants and herbicides, from the impacts of global change. None of these natural and human impacts occurs in complete independence from the others.

The ramifications of the aforementioned massive spruce kill-off in Alaska provide an example of the complexity of ecosystem changes related to global change. Organisms that depend on spruce forests as a habitat are of course affected. Some grazing wildlife may benefit because of the formation of patches of grass as the forest becomes more open. The potential for forest fires is greatly increased by the wood debris on the forest floor, and local lakes and streams have larger volumes of water because the dead trees are no longer losing water to the atmosphere by transpiration.

Some plant species have been expanding their range in response to global warming. In a very general way, long-lived bushes and trees have tended to stay put so far, but short-lived plants such as herbs and various annuals have moved both north (in the Northern Hemisphere) and to higher elevations. Since all plants don't respond to climate change in the same way and at the same rate, the ecosystem must change, affecting all living organisms in the system.

Birds

Birds have faced a lot of hazards during the last century, including loss of habitat, hunting, the use of DDT, and the introduction of competing invasive species. Now global change must be added to the list of perils facing the world's birds. The *2010 State of the Birds Report*, the U.S. Fish and Wildlife Service, and several environmental groups have summarized the risks that global change will create for birds. The sixty-seven species of oceanic birds (albatrosses, petrels) are at greatest global risk, because they reproduce slowly and face problems from a rising, warming ocean. Birds in coastal regions (salt marsh sparrows, plovers), arctic and alpine regions (White Tailed Ptarmigan, Black Turnstone), and grasslands (grouse), as well as land birds from islands (Puerto Rican Parrot), face an intermediate level of risk. Birds from arid lands (Gilded Flicker, Costas Hummingbird), wetlands (various waterfowl), and forests (warblers, flycatchers) are least vulnerable to climate change.

The problems for migratory birds (swifts, nightjars) include the timing of food appearance (e.g., insects) along their routes and at their destinations, and changes in environmental indicators that tell them when to start and when to end migration.

Other Animals

The Sundarbans mangrove forest, the largest in the world, will be largely wiped out by a two-foot sea level rise. Owned by both India and Bangladesh, this is the home of the only Bengal Tiger adapted for life in a mangrove forest. It is boxed in by extensive development around its upland edges, making inland migration of the forest apace with the sea level rise unlikely. The tiger is of course an iconic animal like the polar bear, but the Sundarbans also has 50 reptile species, 300 bird species, 45 types of mammals, and 120 edible fish species. These too will disappear with the rising sea, and some are found only in the Sundarbans.

As with birds and plants, the impact of global change on other animals including mammals, amphibians, insects, and fish is a complex mixture of prey and predators, invasive species, food supplies, timing of food supplies, changing ecosystems, changing habitats, rising temperatures, and changing rainfall patterns, all mixed in with the impact of human activities. The Monarch butterfly, known for its awe-inspiring migration pattern from central Mexico to the eastern United States and Canada, is particularly vulnerable to global change. The forests of central Mexico where the butterfly winters are expected to become wetter and possibly cooler in the next fifty years. The eastern United States and Canada may become warmer and drier and move the optimum habitats further north, extending the required migration distances.

One impact of warming can be measured by the body size of animals. For example, arctic foxes are becoming significantly smaller, according to a study by Yoram Yom-Tov of Tel Aviv University. Smaller bodies allow mammals to endure warmer weather more successfully. Another impact related to sea level rise is the distribution of freshwater fish at the heads of estuaries. In a number of locations reports from fishermen indicate that the range of the fish is decreasing as intruding seawater pushes them upstream.

It is a long and fascinating story with a largely unpredictable ending. Some of the wildlife species which seem to be strongly affected so far by global warming include caribou, polar bears, arctic foxes, gray wolves, tree swallows, painted turtles, toads, and salmon. Which organisms will ultimately adjust to change and which will fall victim to it remains a question to be answered in the coming decades.

Myths, Misinterpretations, and Misunderstandings of the Deniers

MYTH: *Plants are not as sensitive to temperature control as global change models assume, as indicated by numerous examples of "out-of-place" plant assemblages.* These "disjunct" plant assemblages of cold climate plants existing in warm climates are common in the southern United States. One example is the fifty-one species of alpine tundra plants high on San Francisco Mountain, Arizona. Another is the occurrence of Arctic Alpine species found in deep gorges in the Cape Breton Highlands of Nova Scotia. But rather than indicating a lack of sensitivity to temperature, these plant assemblages are believed to be tiny, isolated remnants of the ice age forests and grasslands that once covered the region.

MYTH: *Plants and animals have been through this all before.* During the last two million years there have indeed been at least seven major swings in temperature, as indicated by the advances and retreats of glaciers. The problem is that global change is happening at a rate that is much more rapid than any known in the geologic record. Migration and adjustments to the changing food supply patterns of plants and animals often require more time than is available. The impact of global change on plants and animals does not happen in isolation. The widespread human destruction of habitats through deforestation and agriculture is of course greatly adding to the problem.

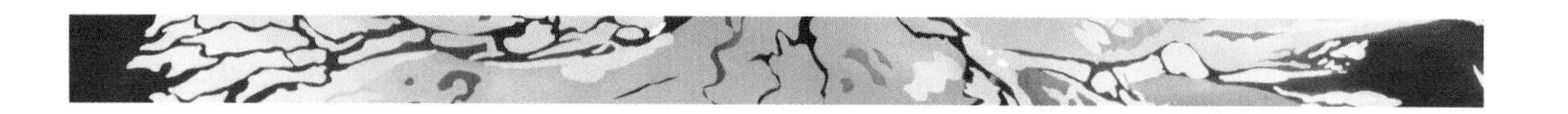

PLAN B

9 ✷ GEOENGINEERING TO THE RESCUE?

The eruption of Mount Pinatubo in the Philippines in 1991 was the second-largest eruption of the twentieth century. Volcanologists successfully warned most of the people living near the mountain, but the massive eruption still managed to kill over seven hundred people, many from roofs collapsing under the pileup of ash. Measurements reveal that the mountain spewed ash twenty-one miles high, spreading vast amounts of aerosol and dust particles into the stratosphere. Mount Pinatubo ejected an estimated seventeen million tons of sulfur dioxide into the stratosphere, a colossal deposit of particles that reduced the amount of sunlight reaching the earth by about 10% (by reflecting sunlight back into space) and resulted in an average cooling of the Earth's surface of nearly 0.6 degree Celsius (1 degree Fahrenheit) in the year following the eruption. The particles from Mount Pinatubo remained in the stratosphere for three years.

As previously discussed, aerosols released by anthropogenic activity can have a cooling effect. This cooling complicates efforts to combat global warming. If we "clean up" pollutants to reduce greenhouse gas emissions, we may experience increased temperatures because we also eliminate aerosols. Such particles contribute to health problems, asthma and lung cancer, but also may cool the atmosphere by reflecting sunlight.

In an essay published in the journal *Climate Change* in 2006, Paul Crutzen, an atmospheric chemist who won the Nobel Prize in chemistry in 1995, proposed injecting sulfates into the stratosphere with balloons or artillery guns to combat "potentially drastic climate heating." Crutzen argued that if "sizeable reductions in greenhouse gas emissions will not happen and temperatures rise rapidly, then climatic engineering . . . is the only option available to rapidly reduce temperature rises and counteract other climatic effects." Crutzen was not the first to propose such a radical idea, but the essay marked a shift in thinking by some in the scientific community, who believed that the time had come to seriously study engineering schemes as a means of quickly counteracting catastrophic climate change. The shift was brought about

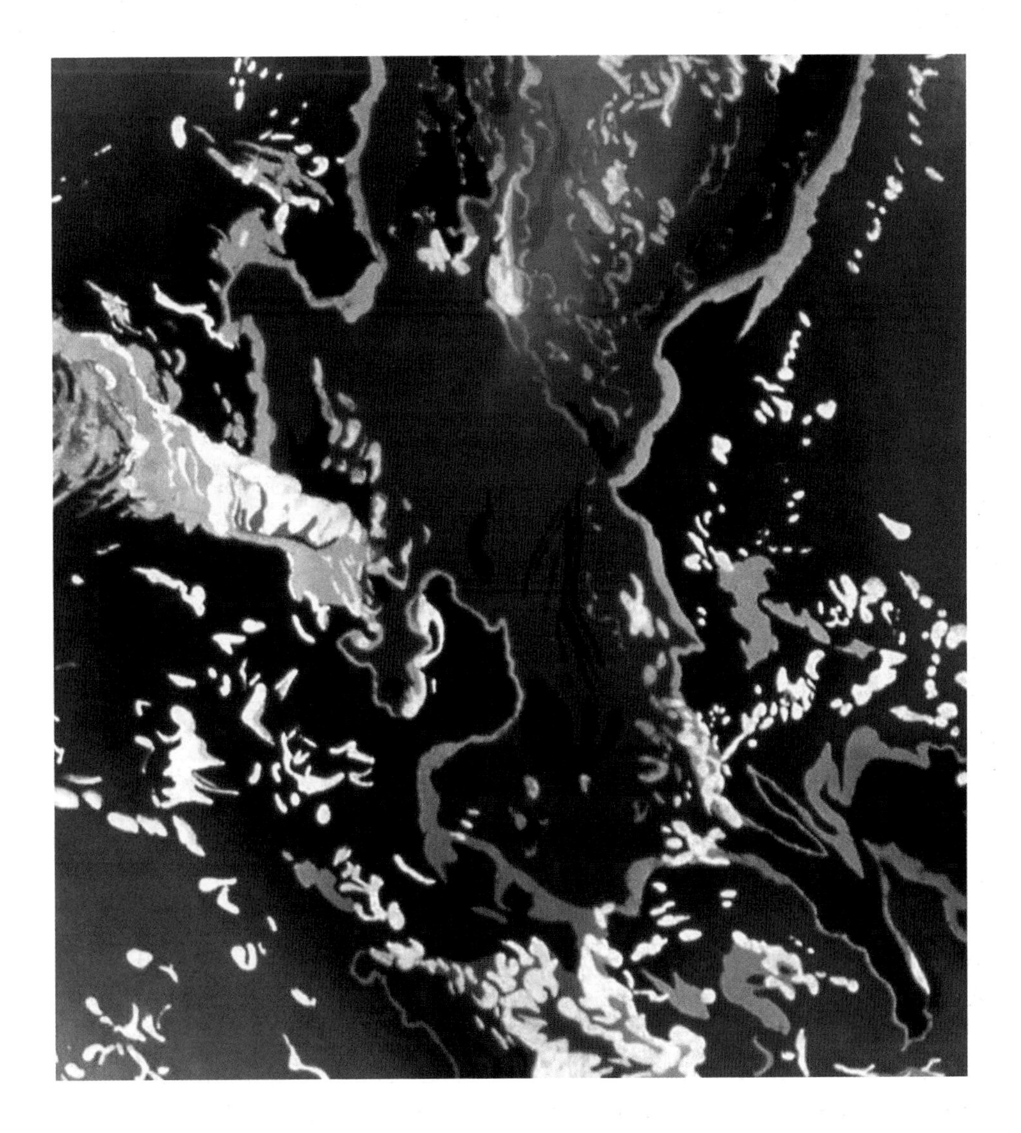

MOUNT PINATUBO, PHILIPPINES

The eruption of Mount Pinatubo in 1991 injected much volcanic ash into the atmosphere and resulted in a cooling of the Earth's atmosphere for more than a year. This event has bolstered the idea of injecting particulate matter or sulfate into the atmosphere to counteract the heating caused by carbon dioxide emissions.

by a concern that we have waited too long and moved too slowly toward decreasing greenhouse emissions and may be on the verge of cataclysmic increases in temperature. Of particular concern is the runaway feedback cycle: Arctic warming reduces surface ice, leading to less sunlight reflection, leading to increased water and surface temperatures, leading to increased release of greenhouse gases including the particularly volatile methane gases from melting permafrost and the ocean floor (methane hydrates), leading to further reductions in surface ice, and so on.

Faced with such dire climate possibilities, science is shifting from thinking about geoengineering schemes to talking about and possibly experimenting with them.

What Is Geoengineering?

Geoengineering is the large-scale manipulation of the environment to counteract anthropogenic climate change. Geoengineering proposals commonly involve two approaches to counteract increased temperatures: greenhouse gas remediation and solar radiation management. Greenhouse gas remediation consists of building devices or manipulating natural systems to lower the levels of CO_2 in the atmosphere and reduce the greenhouse effect. Solar radiation management seeks to offset greenhouse warming by making the Earth more reflective to reduce the incidence and absorption of incoming solar radiation.

GREENHOUSE GAS REMEDIATION

Unlike scrubbers, which capture CO_2 from polluting smokestacks, proposed carbon dioxide removal technology is intended to capture and remove CO_2 from the atmosphere itself, rather than from the source of pollution. One idea is to build machines to capture carbon dioxide from the air and then store it, either deep in the ocean or underground. David Keith, a geophysicist at the University of Calgary, has built a device which successfully removes carbon dioxide and was featured on an episode of the Discovery Channel's "Planet Earth." One of the major problems with using this device is the cost. It requires energy to operate these machines, and the scale of operations would need to be massive to have a significant effect on climate. Optimally, CO_2 removal operations would be located above storage sites, to avoid the added costs of transporting waste. An advantage to these devices is that since they remove carbon dioxide from the ambient air, they would not need to be placed near pollution sources. The underlying strategy differs from the traditional concept of capture and sequestration, in that it promises to capture carbon released from mobile emission sources. In other words, while scrubbers can capture CO_2 emissions from big polluters, such as coal plant smokestacks, Keith's machine would be able to pull carbon directly from the air, potentially negating the impact of transportation-related CO_2 emissions. Keith also proposes combining

CO_2 captured from the air with hydrogen to produce a carbon-neutral transportation fuel, which would have an advantage over conventional biofuels because the production of the fuel would not require the use of land which could otherwise be used to produce food.

There's something admittedly romantic about the notion of building machines to pull CO_2 out of the air. It's the stuff of science fiction, calling to mind the works of Jules Verne. The idea also has an advantage over many other geoengineering schemes in that there are fewer unpredictable or undesirable side effects. For that reason alone it is worthy of significant investigation.

SEEDING THE OCEANS

One of the most extensively studied ideas is to artificially increase planktonic algal blooms. The phytoplankton capture carbon by photosynthesis, acting like a carbon sponge. When they die they carry some of this carbon to the ocean floor. Scientists have discovered that planktonic blooms can be induced by introducing limiting nutrients, such as phosphate or nitrogen, but it appears that one of the most effective ways to create an algal bloom is to fertilize with iron those regions in the sea which otherwise are abundant in phosphate and nitrogen. This has been done successfully a number of times as part of scientific experiments.

Fertilizing the ocean with iron appears to be one of the geoengineering schemes most likely to be employed, for several reasons. First of all, its technological and financial requirements are relatively low: only ships and a supply of soluble iron materials are needed. It is also a particularly attractive option for polluters involved in carbon emissions trading, which is one of the ways countries can meet their climate change mitigation duties under the Kyoto Protocol. Under the protocol, countries can use greenhouse gas removal programs such as reforestation or carbon sinks to meet their emission reduction requirements. Thus if ocean fertilization on a large scale becomes an option for countries looking to meet emission reduction requirements, it could be a highly profitable endeavor. At present, carbon sinks from ocean fertilization are not a tradable emissions commodity, but the possibility of huge profits has already attracted attention from private companies.

Fertilization is not without its detractors. In 2008 the United Nations Convention for Biological Diversity agreed to what the German environment minister Sigmar Gabriel described as a "de facto moratorium." Delegates agreed to refer to the London Convention for guidance on ocean fertilization. In late 2008 the London Convention stated that ocean fertilization activities other than "legitimate scientific research should not be allowed." In January 2009 an Indo-German scientific team sponsored by the Alfred Wegener Institute and others conducted the LOHAFEX experiment in the southwest Atlantic despite protests from environmental groups. The team's research vessel dumped six tons of dissolved iron in a 300-square-kilometer (115-square-mile) patch of the ocean. Results were disappointing, reportedly because a lack of silicic acid resulted in soft-shelled plankton which were consumed by predators, rather than dying off in large numbers and taking carbon with them as they sank to the depths of the ocean floor.

In addition to fertilization using iron, phosphate, and possibly nitrogen, another possible method to create phytoplankton blooms is to increase ocean upwelling to pull nutrients from deep water up to the surface. Like Keith's CO_2 removal machine, this method was featured on Discovery Channel's series "Planet Earth," which showed scientists crafting long plastic pumps operated by wave action. The pumps proved insufficient to withstand the ocean environment, but this idea does offer a potential way to induce plankton blooms without resorting to dumping materials into the ocean.

A report on geoengineering from the Royal Society in September 2009 concluded that ocean fertilization, while potentially cheap, offers only a relatively small capacity to sequester carbon, while noting as well that verifying the carbon sequestration benefit is difficult. The report also cited numerous potential undesirable side effects, including nutrient robbing, in which essential ingredients besides the one being added are removed by the intervention (for instance, nitrogen and phosphate are removed when iron is added), thus depriving downcurrent communities of these ingredients. The report warned that all ocean fertilization proposals involve intentionally changing the marine ecosystem and stressed that the possible consequences are uncertain.

OTHER WAYS TO GO

Other greenhouse gas remediation ideas include reforestation and sequestering carbon dioxide in the form of charcoal, otherwise known as biochar. As previously mentioned, changes in land use account for large amounts of greenhouse gas emissions. In particular, deforestation is a major cause of rising emissions. Forests are a form of carbon sink, and the preservation or planting of trees is particularly appealing as a carbon market commodity—yet another way for countries and businesses to meet their carbon reduction goals. Biochar is produced by taking the carbon dioxide removed from the atmosphere by plants through photosynthesis and then capturing the carbon by burning the organic material in a low-oxygen environment to create charcoal, which is then buried. Instead of burying the biochar one can burn it, providing an alternative to fossil fuels. However, burying biochar reportedly improves soil quality. The Royal Society report questions the efficiency of growing crops for large-scale carbon sequestration and notes that the use of crops to produce renewable biofuels could compete with land used for the production of food. Indeed, this is a serious ethical concern relevant to other forms of biofuels as well.

Solar Radiation Management

The other major geoengineering scheme is solar radiation management, or making the Earth reflect more of the sun's rays back into space. The principal measures aim to increase the Earth's albedo by either brightening the surface or by increasing reflective materials in the atmosphere.

STEVEN CHU'S BRIGHT IDEA

Steven Chu, United States secretary of energy, has proposed painting roofs white and creating white highways to reflect more of the sun's radiation back to space. His comments have sparked the "cool roof" industry, which promises not only to help combat global warming but also to cut down on energy use by keeping houses cooler in the summer. White roofs do not heat up as much as traditional darker roofs, and numerous companies have saved money by resurfacing their roofs in white material. Of course, for there to be any significant cooling effect, resurfacing would have to be done on a grand scale. But Chu's proposal at least represents a relatively simple way to combat warming. The cost of repainting surfaces would be high, but white surfaces could easily be phased in over the next few decades by incorporating them into new structures rather than repainting all existing structures. Also, businesses and homes could realize savings in air-conditioning costs if they opted for a "cool roof" when repairing or replacing their roofs.

Another possibility, which is even cheaper and has the potential to cover a greater area of the Earth's surface, is to plant reflective crops and grasses. Doing this would require caution and could be detrimental to plant and animal diversity if employed on a large scale. Other ideas include covering desert surfaces with a reflective surface, which might be good for reflecting solar radiation but certainly could not be good for desert ecology. Another novel idea is currently being employed in the mountains of Peru, where the inventor Eduardo Gold is painting the top of a mountain peak white, using a mixture of lime, industrial egg white, and water in an effort to cool the local environment and encourage the growth of glaciers, which like most glaciers worldwide are now in retreat.

The potential for cataclysmic climate change has not escaped the attention of Bill Gates, who reportedly has provided $4.5 million to the researchers David Keith and Ken Caldera, two of the biggest names in climate research. The *Vancouver Sun* reports that $300,000 of Gates's money has gone to fund research on another potential solar radiation method of cooling the atmosphere, by increasing cloud condensation nuclei in ocean clouds, in effect brightening the clouds and increasing the solar radiation reflected back toward space. Gates's money is reportedly funding the Silver Lining Project, to determine whether sea water may be sprayed from oceangoing vessels into low-level clouds to accomplish the whitening of marine clouds. Other proposals involve seeding the clouds from airplanes or unmanned drones. The Royal Society points out that this option has the advantage that it could be quickly stopped should unforeseen problems arise and could be strategically employed for the targeted cooling of certain areas.

BACK TO AEROSOLS

As mentioned at the start of this chapter, another solar radiation management idea is to pump aerosols into the stratosphere to reflect sunlight back into space, a process similar to what happens when a powerful volcano explodes and sends dust high into the atmosphere. Unlike the brightening of marine clouds, this approach would have effects on a global scale. Sulfate aerosols can be introduced as gases and would oxidize into particulates. Scientists have proposed pumping these sulfates into the stratosphere with long hoses, high-flying aircraft, or artillery. Like volcanic material, particulates injected into the stratosphere would ostensibly float around for a couple of years, resulting in cooler temperatures. But there could be serious negative ecological side effects, including a possible decrease in the ozone layer, which was one effect observed after the eruption of Mount Pinatubo. In addition, there could be devastating consequences for regional weather. For instance, Luke Oman and his colleagues predict that the African and Asian monsoon seasons could be changed, potentially disrupting the food supply for billions of people. Kevin Trenberth reported that the eruption of Mount Pinatubo disrupted the hydrological cycle and resulted in decreased precipitation.

With such potentially drastic environmental impacts, why on earth would scientists be considering distributing sulfate aerosols into the stratosphere? In part, the answer lies in an understanding among the scientific community, and among climatologists in particular, that there is a potential for cataclysmic climate change, and that the political process may fail to make the policy changes necessary to stave it off. Pumping aerosols into the stratosphere is a comparatively cheap and potentially effective way to cool temperatures worldwide. But this plan, like all the solar radiation management plans, would not decrease the growing ocean acidification problem, nor work to reduce carbon dioxide one iota. Thus, were we to employ the Mount Pinatubo solution as an emergency fix for climate chaos, we would have to continue to place particulates in the stratosphere. If we stopped, the effect would be like that of raising a curtain on a sunny window—the temperature increase would be quick and dramatic.

Geoengineering Embraced by the Fossil Fuels Industry

Throughout this book we have identified various stands taken by those who deny the existence or dangers of climate change. As noted previously, those who stand to benefit from the continued uninhibited burning of fossil fuels may deny the existence of global warming, a stand which is becoming increasingly outlandish given the observable indicators of global climate change, or they may deny that climate change will have disastrous results. Ultimately, the goal of the fossil fuels industry is to delay actions that could hasten a switch to cleaner fuels and result in restrictions on carbon emissions, which would harm their profits. In an essay in April 2009, Alex Steffen, environmental journalist and founder of Worldchanging, argued that the carbon lobby is embracing geoengineering as an argument to delay climate action: "The new climate denialism is all about trying to make the continued burning of fossils fuels seem acceptable, even after the public has come to understand the overwhelming scientific consensus that climate change is real. That's why denialists present geoengineering as an *alternative* to emissions reductions, and couch their arguments in tones of reluctant realism."

Steffen went on to point out that several carbon lobby organizations were touting geoengineering as a viable solution to climate change. Among these were the Cato Institute, the Heartland Institute, and the American Enterprise Institute, which founded the Geoengineering Project to endorse geoengineering fixes.

Some Final Thoughts on Geoengineering

At present geoengineering is in its infancy. Nothing has proceeded beyond the initial research level, and nothing has been applied on a major scale. Geoengineering is not ready to be applied on a major scale, and quite frankly, that's probably a good thing, because the potential for adverse side effects is so high and so unpredictable. While technology and engineering advances will be necessary if we are to move beyond our reliance on fossil fuels and thus curb emissions, geoengineering can only be viewed as an emergency, last-gasp measure. Geoengineering is the stuff of science fiction. If the rosier predictions of its advocates are not borne out, it is easy to envision a dystopian future with artificially darkened skies, acidic oceans, drought, and starvation. Unfortunately, a similarly bleak future is even more likely if we do not cooperate on a global level to reduce greenhouse gas emissions. Geoengineering is more a symptom of the severity of the climate

MORRIS ISLAND LIGHTHOUSE, SOUTH CAROLINA

For more than sixty years this lighthouse in South Carolina has survived storms, rising sea level, and barrier island migration. Constructed originally four hundred meters (one-quarter mile) behind the beach, it now resides four hundred meters seaward of Morris Island. This landmark symbolizes a flexible, non-engineering response to global change.

crisis than a solution. That geoengineering has progressed from an obscure topic into one which is receiving increased research funding and attention in the press should scare all of us. And that climate change is so dire a challenge as to propel some in the scientific community to seriously consider geoengineering should scare us all the more. The anthropogenic climate change crisis that we face requires a political rather than a geoengineering solution.

A Final Word

Today 6.7 billion people populate our tiny planet, and the number inches toward an expected 10 billion by 2050. There are 1,000 people per square kilometer (2,590 per square mile) in Bangladesh, 500 (1,295) in South Korea, 377 (976) in Japan, 360 (932) in India, and 330 (855) in Vietnam. The density of populations places us on a collision course with climate change. Quite simply, we are in the way. We are in the way of storms, sea level rise, shoreline erosion, forest fires, desertification, crop failures, loss of glacial melt water for drinking and agriculture, and all the other expected impacts of global change. And because so many of us require fossil fuels for our chosen mode of existence, we send ever-larger volumes of CO_2 and methane into the atmosphere.

Today there are many economic and political refugees around the world, as well as a smaller number of environmental refugees, such as Pacific Islanders who have fled their flooded homelands, or residents of New Orleans permanently displaced by Katrina's floodwaters. But in the near future much larger numbers of environmental refugees from atolls, deltas, and coastal cities will be added to the list of the displaced. Included in this group will be those who have run out of water or other resources and who also will be seeking relief by moving. The fourteen million people living at elevations below three feet in Bangladesh will be forced to move and find sustenance and shelter in a country that is already one of the world's most densely populated. Right next door is Myanmar, a less densely populated country with potential space for Bangladeshi refugees, but the intervening national boundary may be impermeable.

The potential for conflict is real, and the likelihood is strong that local and regional wars will break out. American and NATO military planners are already contemplating responses to the global warming wars, but one would hope that diplomatic efforts could ward off the conflicts.

The problems can only worsen as global change advances on many fronts, simultaneously with the advancing global population. Of course as long as governments of the world pay only lip service to the problem, we will move no closer to a solution. The carbon industry continues to confound the public with disinformation, delaying the difficult decisions that must be made to

EARTHRISE

"We're not really guilty. We didn't deliberately set out to heat the world," the biologist James Lovelock said in an interview in 2010. But we did, and now the future paths of changes on Earth are uncertain. Now our biggest problem is to convince the deniers to recognize what we've done and where we are heading. When recognition finally occurs, perhaps all of us who occupy this globe can begin to repair our habitat.

significantly reduce greenhouse gas emissions, and weakening any commitment to massive investment in new and sustainable technologies. Those technologies can promise a quality life for the planet's inhabitants without drastically altering the climate. The increasingly serious attention being paid to geoengineering solutions indicates that we may be approaching a point where catastrophic global climate change cannot be avoided. The immediate challenge is for the planet's nations and political leaders to find the courage to put the future of humanity ahead of their own short-term economic interests. Only then will we be able to drastically reduce greenhouse gas emissions and move beyond the fossil fuel age to a future of sustainable energy.

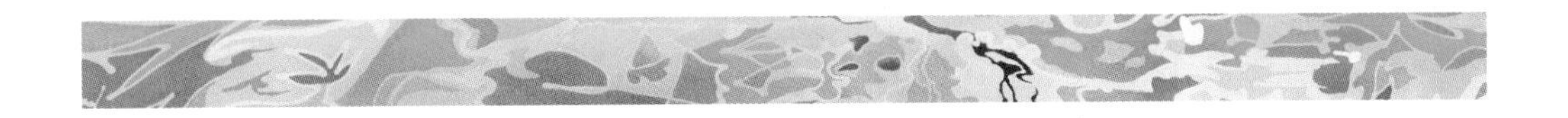

ABOUT THE ART

MARY EDNA FRASER is a master dyer teaching the art of batik worldwide. Her work encompasses the far reaches of the planet. She specializes in the aerial perspective, featuring large-scale art and architectural commissions. For this book she investigates the geography of the dynamic forces of global change, illustrating what Orrin Pilkey thinks is most pertinent.

Batik is a "dye-resist" process in which wax is applied to fabric, creating areas that repel dye, while any unwaxed areas absorb dye. This technique of textile design predates recorded history. Combining modern chemical Proceon dyes, beeswax, paraffin, and silk, Mary Edna's work has a dramatically increased scale and complexity that is unique to the ancient medium. She batiks on silk because it has innate tensile strength as well as an atmospheric quality. Japanese woodblock prints, the impressionist artists, and contemporary painters have been influential in her study of art.

The concept of painting aerial landscapes became Mary Edna's trademark, *Islands from the Sky*, in 1980 when she and her brother were flying in the family's vintage 1946 Ercoupe. A lifelong exposure to flight began at two weeks of age with her Daddy flying to Grandmother's house. Fraser has photographed most of the eastern and western coastlines of the United States and many landscapes abroad, observing both breathtakingly beautiful and disturbing scenes.

She employs a variety of sources to define a sense of place, including on-the-ground watercolors and sketches, satellite imagery, maps, and photographs furnished by scientists of distant regions. Fast film and global interactive mapping have become available during Fraser's thirty-year career. A Nikon D90 digital camera has replaced her 35mm film cameras in recent years. Often renting a plane on location, she hires an instructor and sets up her own shots to the best altitude and vantage point. Usually an aerial research excursion will yield about five hundred photographs, but few are chosen as design candidates.

Mary Edna's methods are uniquely her own. She stretches the design drawn in pencil on silk horizontally between two sawhorses. Working like a painter, she applies up to four layers of wax and color to create depth. The fabric must be waxed on both sides to ensure complete absorption. The tools of the trade include soft-bristle wax brushes and the tjanting, a spouted copper bowl with a handle, which is used to create very fine lines with hot liquid wax. A typical batik, measuring twelve by four feet, may require five pounds of wax. Because she takes care not to bend or fold the waxed silk, the crackling of wax characteristic to most batiks is rarely seen in Mary Edna's refined technique. The process requires care and precision since there is no simple way to erase an unwanted drop of wax or dye.

The dyes come in powder form and must be mixed in exact proportions of water, urea, Calgon, baking soda, and washing soda. Any mistake in chemistry will cause the dyes to bleed, ruining the batik. Mary Edna tests colors on paper and spends several hours working out a color harmony, often comparing dyes to nature's palette. Applying the liquid dyes with a brush allows for subtle color transitions. Ironing out the wax between sheets of newsprint sets the fiber-reactive dyes, making the color permanent.

The slowness of the labor-intensive process encourages a contemplative approach. The batiks convey perspectives that the human eye, maps, and cameras cannot fully reveal. Collaborating with Orrin Pilkey has broadened Fraser's horizons, taking her on adventures to threatened environments. When selecting landscapes to batik, Pilkey and Fraser argue over which images best depict the science. A design is chosen when both partners are pleased with both aesthetics and clarity. With Pilkey's science and Mary Edna's art, they engage the viewer intellectually and emotionally. For more information on the process of batik go to www.maryedna.com.

Earthscaping (p. 3)
47" × 47" 2001
In *Earthscaping* the areas representing ice on top of the planet and white areas were waxed first, and a dye bath of the blues of the water followed. When these areas were dry a second wax resist was applied, and the silk was dyed in green-to-red hues for the land. The third layer of wax covered most of the landscape, and the mountains were then dyed. A final waxing left a few small spaces for the darkest dye bath, to create depth in the mountains and the deep indigo sky.

Slopes of Mauna Loa, Hawaii (p. 6)
184" × 36" 1994
Landsat images reveal the diversity of Earth's landforms and show how the science of remote sensing can capture inaccessible scenes. Dark lava flows run down the slopes of Mauna Loa, and rich, deep red vegetation grows on the island. This batik of the world's largest active volcano was awarded recognition by the U.S. Geological Survey's EROS Data Center in 1998.

Selenga Delta, Russia (p. 8)
46" × 45" 2006
Located in southeast Siberia, Lake Baikal is on the UNESCO World Heritage List and

is the oldest (25 million years) and deepest (1,700 meters, or 5,580 feet) lake in the world. The Selenga River empties into one of the most biologically diverse lakes on Earth. This is my first piece of art using Google Earth as a reference.

Amazon River (p. 12)
103" × 35" 1995
In 1995 I was the artist for the NASA Art Program. Its curator, Bert Ulrich, allowed me access to every photographic image taken from space to date. This elegant slice is based on two tributaries of the Amazon. It shows in yellow the environmental impact of declining forests. Delicate silk fabric evokes the fragility of the Earth, the condition of which NASA continually monitors through environmental studies that employ satellite technology.

Wilsons Promontory, Australia (p. 19)
27.25" × 35.5" 2009
The southernmost point of the Australian mainland contains the largest coastal wilderness area in Victoria. This is one of four hundred fires recorded on 7 February 2009 (known as Black Saturday) which burned out of control. The massive bushfire burned over five weeks and destroyed close to 50% of the National Park.

Murray River, Australia (p. 21)
31" × 27.75" 2009
A flight to photograph the Murray River in 2007 took me over barren land connected by brightly colored irrigation ditches. A damask silk with ginkgo leaves in the pattern was used to illustrate a small segment of Australia's longest river. Like the ginkgo, the river is a living relic of our ancient world.

St. Louis, 1988 and 1993 (p. 23)
35.5" × 21" 2007
Marty and Cathy Wice saw my work at an exhibition in St. Louis. They had experienced the devastating flooding of the Mississippi and Missouri rivers and commissioned this work as well as another. I flew over the site in a jet and spent time on the ground by the rivers' edges. The two satellite images from 1988 and 1993 clearly illustrate the flood; the city is indicated in red.

Hurricane Katrina (p. 24)
51" × 54" 2007
Hurricane Katrina was a category 4 storm when it made landfall. As the levee system built by the U.S. Army Corps of Engineers catastrophically failed, the destruction brought widespread criticism of government reactions. The costliest natural disaster in United States history at the time still echoes in human torment. This batik is based on a NOAA satellite image taken before Katrina struck the coast.

Gulf Oil Spill (p. 26)
55" × 35" 2010
During eleven flights conducted between 6 and 25 May at the request of the National Oceanic and Atmospheric Administration, NASA's Airborne Visible/Infrared Imaging Spectrometer (AVIRIS) extensively mapped

the region affected by the BP oil rig disaster on 17 May 2010. Crude oil on the surface appears orange to brown. The white on the batik represents boats and platforms, which seem insignificant to the scale of the damage.

Pacific Full Moon (p. 30)
50" × 44" 1999
Two hundred miles above Earth, the crew of the space shuttle *Columbia* photographed the full moon setting over the Pacific Ocean. Decades of photography offer new avenues for scientific study of our ever-changing planet. One of the astronauts, Jay Apt, took the photograph on which this batik is based, using a modified Hasselblad medium-format camera with a 100 mm lens and Kodak Ektachrome 64 professional film. It was fascinating to talk with Jay on the phone about his otherworldly experiences.

Iceberg (p. 35)
77" × 44.5" 2008
The iceberg batik is an artistic rendering from a digital composite photograph by Ralph A. Clevenger, who is on faculty at Brooks Institute. The portion above water depicts an Antarctic iceberg weighing approximately 300 million tons; the underwater image was actually shot above water in Alaska and flipped upside down. The art accurately represents the amount of an iceberg that is hidden below the surface. Only one-seventh to one-eighth of an iceberg can be seen above water, thus the phrase "the tip of the iceberg."

Global Perception (p. 47)
876 square feet, detail 1999
This is a segment of the largest batik sculpture in the world to date, designed for a four-story trapezoid atrium. The depiction of the Earth in silk, measuring 21 by 9 feet, is an adaptation of Buckminster Fuller's Projection Dymaxion Map (1930). The first flat map of the entire surface of the Earth, it reveals our planet as one island in one ocean, without any visually obvious distortion of the relative shapes and sizes of the land areas, and without splitting any continents.

Moulin, Greenland (p. 56)
54" × 36" 2010
The famous photograph inspiring this massive moulin is credited to Roger J. Braithwaite at the University of Manchester. The almost vertical plunge shows millions of gallons of water cascading to bedrock. I eliminated the human beings standing on the edge, which were tiny in this icy landscape.

Glacial Canyon, Alaska (p. 59)
60" × 36" 1993
Glaciers still shape the Alaskan landscape, carving deep fjords as they push huge quantities of rock and earth seaward. My friend Dana Bell took the reference photograph for this batik of the Triumvirate Glacier northwest of Anchorage. Layers of dyes and the striped silk help to reveal the power of the ancient ice that sculpted this canyon.

Mount McKinley, Alaska (p. 60)
69" × 46" 1993
The highest mountain on the North American continent, Mount McKinley, is in Alaska's Denali Mountain Range. The batik is based on an aerial photo by William Wilson and was my first challenge in depicting ice on silk. It is interesting that a landscape so far away from my habitat will influence my children's future.

Bhutan's Himalayas (p. 62)
37.5" × 54" 2008
In July 2008 I was invited to demonstrate batik on the National Mall in Washington for NASA at the Smithsonian Folklife Festival. The tiny Asian nation of Bhutan was also featured. I penciled onto silk a satellite image taken on 20 November 2001 that was used to study glaciers. Bhutanese in their indigenous dress loved seeing their home becoming art.

Kilimanjaro, Africa (p. 64)
34" × 44.5" 2008
As the tallest freestanding mountain rise in the world, Kilimanjaro has seen its recent ice cap volume drop by more than 80%. In this silken aerial view, clouds shroud the iconic mountain. This batik stands out in its peaceful nature.

Charleston Airborne Flooded, South Carolina (p. 70)
97" × 35" 2010
Matt Pendleton, a spatial analyst with the IM Systems Group working for NOAA Coastal Services Center in Charleston, prepared a map which shows a projected rise in sea level of 1.4 meters (4.6 feet) by the year 2100. This batik illustrates a mid-range projection, flooding the present day shoreline of my home city.

Spencer Gulf, Australia (p. 71)
26" × 37" 2009
Located on the southeastern coast of Australia, Spencer Gulf is a bay the size of the Chesapeake Bay, bordered by the Indian Ocean. My task for this piece of art was to make the barrier islands visible, drawing attention to the past in the modern landscape. The tjanting tool was employed liberally to make bold, white lines, leaving the mark of the human hand in the final artwork.

Pteropod (Limacina helicina) (p. 75)
18" × 14" 2010
Dr. Russ Hopcroft at the Institute of Marine Science, University of Alaska, Fairbanks, provided the reference photograph for this work, taken in the Arctic. This gorgeous, minuscule creature has two paddle-like wings used to row through the water, hence the common name "sea-butterflies."

Northwest Passage (p. 79)
38.5" × 54" 2008
In the summer of 2007 NASA's Aqua satellite imagery indicated that sea ice was retreating to a sufficient extent to change the character of the Northwest Passage. Amid the Canadian Arctic Archipelago, the waterways connecting the Atlantic and Pacific oceans emulate an Old World map in this batik.

Boston II, Massachusetts (p. 85)
43" × 98.5" 2007
Using my artistic license, I have removed the visual noise of human development of Boston's islands seen in the foreground, bringing the harbor back to its natural state. Each city has an individualized dynamic skyline as well as shoreline.

Edingsville Beach, South Carolina (p. 86)
79.5" × 35" 2009
Flying in the family's vintage 1946 Ercoupe with my brother Burke is an invigorating way to take aerial photographs. In 1983 my Nikon 35 mm film camera caught this image. Orrin Pilkey recognized the beach as the former Edingsville location and requested the art. I would consider this scene my aerial backyard.

Shishmaref's Shores, Alaska (p. 88)
34" × 98" 1996
This piece is taken from a high-elevation black-and-white photograph, which I rotated 180 degrees, as if flying toward the mainland. The barrier islands became totemic canoes sailing through dark waters. The wax was cracked to age the piece and the dark final dyes are the color of a bruise, evoking the Inuit population's pain of possibly losing their ancestral home.

Atafu Atoll, Tokelau, South Pacific (p. 90)
22" × 22" 2010
The satellite picture of the Atafu Atoll, taken on 2 March 2009, was the image of the day on NASA's Earth Observatory page. The atoll, in the southern Pacific Ocean, is a territory of New Zealand. The typical ring shape of the atoll is the result of coral reefs building up around a volcanic island. Over geologic time the central volcano subsided, leaving the fringing reefs and central submerged lagoon corals. Orrin Pilkey and I looked for hours before we agreed on an atoll for me to use, one where a rising sea level makes future habitability uncertain.

South of Ocracoke, North Carolina (p. 92)
36" × 36" 2001
My father flew me on an all-day adventure to the Outer Banks from Fayetteville, North Carolina, in our family plane. With the cockpit open in thirty-five-knot winds, I photographed specific barrier islands requested by Orrin Pilkey. On the horizon, mainland had disappeared because the islands are so far out to sea.

Bangladesh (p. 95)
46" × 43" 2010
On 26 March 2009 Earth Snapshot, a daily view of the planet on the web, featured the Ganges and Jamuna rivers flowing into the Bay of Bengal. The wide rivers visible to the north, below the Himalayas, are loaded with sediment, which creates the multiple winding paths on the silk. This new color palette depicts flesh tones, vegetation, and water in an intricate design.

Mouths of the Mekong, Vietnam (p. 96)
52" × 47" 1998
Matt Stutz found a military map at the Library of Congress that became the basis of this batik. The bloody color of the South China Sea contrasts with the edges of the tropical forests and resembles a sorrowful, skeleton-faced Buddha. Land mines from the Vietnam War still threaten those who remain in this countryside.

Laguna Madre, Mexico (p. 99)
33" × 35.75" 2009
For this artwork Orrin Pilkey supplied the aerial photograph. I enjoyed the quiet, serene feeling that the site evoked. Like most batiks, this one took a month to complete. The batik process is detailed and intensive, yet the meditative quality of the medium makes for a calming experience.

Great Barrier Reef II, Australia (p. 100)
104" × 45" 2008
My first snorkeling adventure, in 2007 on the Great Barrier Reef, was my initiation into underwater photography. This batik is a synthesis of that experience, which was like flying under water, with colors and shapes constantly in motion. This aquatic excursion makes me want to explore this realm in different parts of the world.

Sinking Colombian Shores (p. 101)
34" × 63" 1998
When Orrin Pilkey was working on a grant from the National Geographic Society in 1995 to study the Colombian barrier islands, he came to my studio with the original maps of the newly discovered islands. It was astonishing to me to hold these papers in my hands, knowing that even today new knowledge is coming to light about our planet, and that I would pioneer the illustration of the islands.

Tsunami (p. 103)
38" × 35" 2005
According to the U.S. Geological Survey, the earthquake that triggered the great tsunami of 2004 released energy equivalent to 23,000 atomic bombs as powerful as the one dropped on Hiroshima. I used the NOAA animation of the tsunami in Indonesia (Sumatra) on 26 December 2004 as a reference. The light blue in the ocean shows the path of the destruction.

Yukon Delta, Alaska (p. ii, 105)
44" × 44" 2006
In southwest Alaska the waters of the Yukon and Kuskokwim rivers flow through a vast treeless plain, or tundra. Waterfowl, shorebirds, and fish inhabit this remote refuge, as do nearly 25,000 Yupik Eskimo, who rely on hunting and fishing for food. The delta resembles an upside-down flower in full bloom.

Flying the Everglades, Florida (p. 106)
96" × 36" 1993
My brother and I rented a plane in the Keys and began a photographic journey of the Ev-

erglades in blue skies. The aerial landscape here is ethereal and unique. My eyes were busy viewing through the lens of the camera. When I looked out I saw a brooding gray sky, and a tiny blue passage in the clouds allowed us to escape the bad weather that had blown in rapidly. We landed in the wind safely, like a crab, sideways.

Mount Pinatubo, Philippines (p. 111)
31" × 36" 2001
The eruption of Mount Pinatubo produced beautiful sunsets and sunrises all around the globe for years. The cooling effect of the volcanic ash in the atmosphere has been used as evidence to support the plausibility of geoengineering. My friend Jeff Kopish commissioned this piece after working for the Peace Corps in this location.

Morris Island Lighthouse, South Carolina (p. 118)
44" × 36" 2001
I enjoy every part of making art, from the leap in my heart when I see an amazing shot through the camera lens to the final batik. My intent is to convey the essence of place. Once surrounded by land, the Morris Island Lighthouse guided ships into Charleston Harbor.

Earthrise (p. 120)
143" × 36" 2003
Earth rising as seen from the moon, taken by Apollo 11 in 1969, is a thought-provoking image. Seeing our moon's surface in the foreground, with our planet beyond, brings to mind the fragility of our home.

UNLIKE THE OSTRICH with its head in the sand, humankind can face the challenges of climate change. Scientific fact must inspire ingenuity and prod our government to initiate more effective legislation to improve the future of the stressed planet we call home.

BIBLIOGRAPHY

Alfred Wegener Institute. "Lohafex Provides New Insights on Plankton Ecology: Only Small Amounts of Atmospheric Carbon Dioxide Fixed." 23 March 2009, www.awi.de/en/news/press_releases/detail/item/lohafex_provides_new_insights_on_plankton_ecology_only_small_amounts_of_atmospheric_carbon_dioxide/?cHash=1c5720d7a1.

Allègre, C, and D. de Montvallon. *L'imposture climatique: ou la fausse écologie*. Paris: Plon, 2010.

American Friends of Tel Aviv University. "Animals Cope with Climate Change at the Dinner Table: Birds, Foxes and Small Mammals Adapt Their Diets to Global Warming." Science Daily, 11 February 2010, www.sciencedaily.com/releases/2010/02/100209152235.htm, retrieved 4 June 2010.

Beck, C. "Water Vapor Is Indeed a Powerful Greenhouse Gas, but There Is Plenty of Room for CO_2 to Play a Role; Part of the How to Talk to a Climate Skeptic Guide: Responses to the Most Common Skeptical Arguments on Global Warming." 25 December 2006, www.grist.org/article/water-vapor-accounts-for-almost-all-of-the-greenhouse-effect.

Beck, G., and K. Balfe. *An Inconvenient Book: Real Solutions to the World's Biggest Problems*. New York: Simon and Schuster, 2007.

Bellamy, D. "Letter: Glaciers Are Cool." *New Scientist*, 16 April 2005.

Blockstein, D. E., and L. Wiegman. *The Climate Solutions Consensus*. Washington: Island, 2010.

Bloetscher, F., D. E. Meeroff, and B. N. Heimlich. "Improving the Resilience of a Municipal Water Utility against the Likely Impacts of Climate Change: A Case Study: City of Pompano Beach Water Utility." Boca Raton: Florida Atlantic University, 2009.

Booker, C. *The Real Global Warming Disaster: Is the Obsession with "Climate Change" Turning Out to Be the Most Costly Scientific Blunder in History?* London: Continuum, 2009.

Boykoff, J., and M. Boykoff. "Balance as Bias: Global Warming and the U.S. Prestige Press." *Global Environmental Change* 14 (2004), 125–36, www.eci.ox.ac.uk/publications/downloads/boykoff04-gec.pdf.

Broecker, W. S., and R. Kunzig. *Fixing Climate: What Past Climate Changes Reveal about the Current Threat—and How to Counter It*. New York: Hill and Wang, 2008.

Brown & Williamson. 1969. Smoking and health proposal, tobaccodocuments.org/landman/332506.html, document codes 690010951–59.

Bush, D. M., O. H. Pilkey, and W. J. Neal. *Living by the Rules of the Sea*. Durham: Duke University Press, 1996.

Caldeira, K., and M. E. Wickett. "Oceanography: Anthropogenic Carbon and Ocean pH." *Nature* 425 (2003), 365, doi:10.1038/425365a.

Cao, L., K. Caldeira, and A. K. Jain. "Effects of Carbon Dioxide and Climate Change on Ocean Acidification and Carbonate Mineral Saturation." *Geophysical Research Letters* 34 (2007), L05607, doi:10.1029/2006GL028605.

Carnegie Institution. "Carbon Dioxide's Effects on Plants Increase Global Warming, Study Finds." ScienceDaily, 19 June 2010, www.sciencedaily.com/releases/2010/05/100503161435.htm, retrieved 19 June 2010.

Chambers, M. 30 May 2008. "U.N. Talks Halt Plans for Oceans Absorb CO_2." Reuters, www.reuters.com/article/idUSL2981194420080530.

Chen, X., and Y. Zong. "Major Impacts of Sea Level Rise on Agriculture in the Yangtze Delta Area around Shanghai." *Applied Geography* 19 (1999), 69–84.

"The Clouds of Unknowing." *Economist*, 20 March 2010, 83–86.

Collyns, Dan. "Can Painting a Mountain Restore a Glacier?" BBC News, 17 June 2010, www.bbc.co.uk/news/10333304.

Connor, Steve. "Obama's Climate Guru: Paint Your Roof White!" *Independent*, 27 May 2009, www.independent.co.uk/environment/climate-change/obamas-climate-guru-paint-your-roof-white-1691209.html.

Curry, J. A., P. J. Webster, and G. J. Holland. "Mixing Politics and Science in Testing the Hypothesis That Greenhouse Warming Is Causing a Global Increase in Hurricane Intensity." *Bulletin of the American Meteorological Society* 87, no. 8 (2006), 1025–37.

Denton, G. H., et al. "The Last Glacial Termination." *Science* 328 (2010), 1652–56, doi:10.1126/science.1184119.

Doney, S. C., W. M. Balch, V. J. Fabry, and R. A. Feely. "Ocean Acidification: A Critical Emerging Problem for the Ocean Sciences." *Oceanography* 22, no. 4 (2009) [special issue: "The Future of Ocean Biogeochemistry in a High-CO_2 World"].

Doran, P. T., and M. K. Zimmerman. "Examining the Scientific Consensus on Climate Change." *Climate Change* 90, no. 3 (2009).

Draper, R. "Australia's Dry Run." *National Geographic*, April 2009, 34–59.

Dyck, M. G., W. Soon, R. K. Baydack, et al. "Polar Bears of Western Hudson Bay and Climate Change: Are Warming Spring Air Temperatures the 'Ultimate' Survival Control Factor?" *Ecological Complexity* 4 (2007), 73–84, doi:10.1016/j.ecocom.2007.03.002.

———. "Reply to Response to Dyck et al. (2007) on Polar Bears and Climate Change in Western Hudson Bay by Stirling et al." *Ecological Complexity* 5 (2008), 289–302, doi:10.1016/j.ecocom.2008.05.004.

Dyer, Gwynne. "*Climate Wars: The Fight for Survival as the World Overheats*:" Oxford: One World. 2010.

Environmental Defense Fund. "Warming and Wildlife: Monarch Butterfly: Vibrant Fliers at Risk." www.edf.org/page.cfm?tagID=42763, retrieved 4 June 2010.

Ericson, J. P., et al. "Effective Sea Level Rise and Deltas: Causes of Change and Human Dimension Implications." *Global and Planetary Science* 50 (2006), 63–82.

European Science Foundation. "Leading Scientists Call for More Effort in Tackling Rising Ocean Acidity." ScienceDaily, 19 May 2010, www.sciencedaily.com/releases/2010/05/100519081444.htm, retrieved 26 May 2010.

"Europeans' Attitudes towards Climate Change." Special Eurobarometer 32, TNS Opinion and Social, ec.europa.eu/public_opinion/archives/ebs/ebs_322_en.pdf.

Fiala, N. "How Meat Contributes to Global Warming." *Scientific American*, February 2009, 72–75.

Friel, H. *The Lomborg Deception: Setting the Record Straight about Global Warming*. New Haven: Yale University Press, 2010.

Gaskill, Alvia. "Desert Area Coverage, Global Albedo Enhancement Project." 2004, www.global-warming-geo-engineering.org/Albedo-Enhancement/Surface-Albedo-Enhancement/Calculation-of-Coverage-Areas-to-Achieve-Desired-Level-of-ForcingOffsets/Desert-Area-Coverage/ag28.html.

"Geoengineering the Climate: Science, Governance and Uncertainty." Royal Society, 1 September 2009, royalsociety.org/geoengineering-the-climate.

Geological Society of London. "Acidifying Oceans Spell Bleak Marine Biological Future 'by End of Century,' Mediterranean Research Finds." ScienceDaily, 25 August 2010, www.sciencedaily.com/releases/2010/08/100825093651.htm, retrieved 26 August 2010.

Getter, K. L., D. B. Rowe, G. P. Robertson, et al. "Carbon Sequestration Potential of Extensive Green Roofs." *Environmental Science and Technology* 43, no. 19 (2009), 7564–70, doi:10.1021/es901539x.

Giddens, A. *The Politics of Climate Change*. Cambridge: Polity, 2009.

Gladwell, M. *The Tipping Point: How Little Things Can Make a Difference*. Boston: Little, Brown, 2000.

Goodell, J. *How to Cool the Planet: Geoengineering and the Audacious Quest to Fix Earth's Climate*. New York: Houghton Mifflin Harcourt, 2010.

Goreham, S. *Climatism! Science, Common Sense and the 21st Century's Hottest Topic*. New Lenox, Ill.: New Lenox, 2010.

Greenpeace. "Koch Industries Secretly Funding the Climate Denial Machine." 2010, www.greenpeace.org/usa/press-center/reports4/koch-industries-secretly-fund.

Hansen, J. E. 2007: "Scientific Reticence and Sea Level Rise." *Environmental Research Letters* 2 024002, 2007, iopscience.iop.org/1748-9326/2/2/024002/fulltext, retrieved 22 June 2010.

Hansen, J. *Storms of My Grandchildren: The Truth about the Coming Climate Catastrophe and Our Last Chance to Save Humanity*. New York: Bloomsbury, 2009.

Harkinson, J. "Climate Change Deniers without Borders." *Mother Jones*, December 2010.

Harris, N. R. P., et al. "Trends in Stratospheric and Free Tropospheric Ozone." *Journal of Geophysical Research* 102, no. D1 (1997), 1571–90, doi:10.1029/96JD02440.

Heimlich, B. N., F. Bloetscher, D. E. Meeroff, and J. Murley. "Southeast Florida's Resilient Water Resources: Adaptation to Sea Level Rise and Other Climate Change Impacts." Boca Raton: Florida Atlantic University, 2009.

Herberger, M., et al. "The Impacts of Sea Level Rise on the California Coast." Sacramento: California Climate Change Research Center, 2009.

Hoffmann, G. "Claude Allègre: The Climate Imposter." RealClimate.org, http//www.realclimate.org/index.php/archives/2010/04/claude-allegre-the-climate-imposter/ 2010.

Hoggan, J., and R. Littlemore. *Climate Cover-up: The Crusade to Deny Global Warming*. Vancouver, B.C.: Greystone, 2009.

Holsten, E. H., R. W. Thier, A. S. Munson, and K. E. Gibson. "The Spruce Beetle." U.S. Department of Agriculture Forest Service, Forest Insect and Disease Leaflet 127 (1999), www.na.fs.fed.us/spfo/pubs/fidls/sprucebeetle/sprucebeetle.htm.

Houghton, J. *Global Warming: The Complete Briefing*, 4th edn. Cambridge: Cambridge University Press, 2009.

Hughes, T. J. "The Weak Underbelly of the West Antarctic Ice Sheet." *Journal of Glaciology* 27 (1981), 518–25.

Hulme, M. *Why We Disagree about Climate Change: Understanding Controversy, Inaction and Opportunity*. Cambridge: Cambridge University Press, 2009.

IAHS/UNESCO. "Fluctuations of Glaciers, 1990–1995," vol. 7. Zurich: World Glacier Monitoring Service, 1998.

Institut de France, Académie des Sciences. Le Changement Climatique, http://media.enseignementsup- recherche.gouv.fr/file/2010/35/0/Changement_climatique_octobre_2010_159350.pdf.

Intergovernmental Panel on Climate Change. "Summary for Policymakers." *Climate Change 2007: The Physical Science Basis*. Cambridge: Cambridge University Press, 2007.

IPCC. *Special Report on Emissions Scenarios*. Cambridge: Cambridge University Press, 2000.

Keith, D. W. "Why Capture CO_2 from the Atmosphere?" *Science* 325 (2009), 1654–55.

Kelley, J. T., O. H. Pilkey, and J. A. G. Cooper, eds. *America's Most Vulnerable Coastal*

Communities. Boulder: Geological Society of America, 2009 [Geological Society of America Special Publication 460].

Kennedy, C., J. Steinberger, B. Gasson, et al. "Greenhouse Gas Emissions from Global Cities." *Environmental Science and Technology* 43, no. 19 (2009), 7297–7302, doi:10.1021/es900213p.

Kintisch, E. *Hack the Planet: Science's Best Hope—or Worst Nightmare—for Averting Climate Catastrophe*. Hoboken, N.J.: John Wiley and Sons, 2010.

Kitcher, P. "The Climate Change Debates." Sciencexpress, 27 May 2010, 10.1126/science.1189312.

Kleypas, J. A., and K. K. Yates. "Coral Reefs and Ocean Acidification." *Oceanography* 22, no. 4 (2010) [special issue: "The Future of Ocean Biogeochemistry in a High-CO_2 World"].

Knutson, T. R., J. L. McBride, J. Chan, et al. "Tropical Cyclones and Climate Change: Review." *Nature Geoscience* 3 (2010), 157–63, doi:10.1038/ngeo779.

Kolbert, E. *Field Notes from a Catastrophe: Man, Nature, and Climate Change*. New York: Bloomsbury, 2007.

———. "Changing Rains." *National Geographic*, April 2009, 60–65.

———. "Up in the Air." *New Yorker*, 12 April 2010, 21–22.

Kump, L. R., T. J. Bralower, and A. Ridgwell. "Ocean Acidification in Deep Time." *Oceanography* 22, no. 4 (2009), 94–107 [special issue: "The Future of Ocean Biogeochemistry in a High-CO_2 World"].

Krupp, F., and M. Horn. *Earth: The Sequel: The Race to Reinvent Energy and Stop Global Warming*. New York: W. W. Norton, 2008.

Kwok, R., and D. A. Rothrock. "Decline in Arctic Sea Ice Thickness from Submarine and ICESat Records: 1958–2008." *Geophysical Research Letters* 36 (2009), L15501, doi:10.1029/2009GL039035.

Lomborg, B. *The Skeptical Environmentalist: Measuring the Real State of the World*. Cambridge: Cambridge University Press, 2001.

———. *Cool It: The Skeptical Environmentalist's Guide to Global Warming*. Vintage, 2008.

———. *Smart Solutions to Climate Change*. Cambridge: Cambridge University Press, 2010.

Loster, T. "Flood Trends and Global Change." *Proceedings of the EuroConference on Global Change and Catastrophe Risk Management: Flood Risk in Europe*. Laxenburg, Austria: IIASA, 1999.

Lovelock, J. *The Revenge of GAIA: Earth's Climate Crisis and the Fate of Humanity*. New York: Basic Books, 2006.

———. "James Lovelock on the Value of Sceptics and Why Copenhagen Was Doomed." Transcript of interview by Leo Hickman of the *Guardian*, 2010, www.guardian.co.uk/environment/blog/2010/mar/29/james-lovelock.

Loucks, C., S. Barber-Meyer, Md. A. A. Hossain, et al. 2010: "Sea Level Rise and Tigers: Predicted Impacts to Bangladesh's Sundarbans Mangroves: A Letter." *Climatic Change* 98 (2010), 291–98, doi:10.1007/s10584-009-9761-5.

Luo, Y., B. Su, W. S. Currie, D. R. Zak, et al. "Progressive Nitrogen Limitation of Ecosystem Responses to Rising Atmospheric Carbon Dioxide." *BioScience* 54, no. 8 (2004), 731–39, doi:10.1641/0006-3568(2004)054(0731:PNLOER)2.0.CO;2.

Lyman, J. M., S. A. Good, V. V. Gouretski, et al. "Robust Warming of the Global Upper Ocean. *Nature* 465 (20 May 2010), 334–37, doi:10.1038/nature09043.

Mann, M. E., and L. R. Kump. *Dire Predictions: Understanding Global Warming: The Illustrated Guide to the Findings of the IPCC, Intergovernmental Panel on Climate Change*. New York: DK, 2008.

Mann, M. E., et al. "Global Signatures and Dynamical Origins of the Little Ice Age and Medieval Climate Anomaly." *Science* 326 (2009), 1256–60, doi:10.1126/science.1177303.

May, E., and Z. Caron. *Global Warming for Dummies*. Missisauga, Ont.: John Wiley and Sons, 2009.

McCandless, D., and H. L. Williams. "Climate Consensus? How Many US Scientists Disagree with Human-Induced Climate Change?" 2009, www.informationisbeautiful.net/2009/climate-change-a-consensus-among-scientists/comment-page-1.

Mercer, J. H. "Antarctic Ice and Sangamon Sea Level." International Association of Scientific Hydrology, Commission of Snow and Ice, General Assembly of Bern, Publ. no. 79 (1968), 217–25.

Michaels, D. *Doubt Is Their Product: How Industry's Assault on Science Threatens Your Health*. New York: Oxford University Press, 2008.

Michaels, P. J. *Meltdown: The Predictable Distortion of Global Warming by Scientists, Politicians, and the Media*. Washington: Cato Institute, 2004.

Michaels, P. J., and R. C. Balling. "Climate of Extremes: Global Warming Science They Don't Want You to Know." Washington: Cato Institute, 2009.

Monbiot, G. "Pretending the Climate E-mail Leak Isn't a Crisis Won't Make It Go Away." 2009, www.guardian.co.uk/environment/georgemonbiot/2009/nov/25/monbiot-climate-leak-crisis-response.

Montford, A. W. *The Hockey Stick Illusion: Climategate and the Corruption of Science (Independent Minds)*. London: Stacey International, 2010.

Moore, T. C. "Origin and Disjunction of the Alpine Tundra Flora on San Francisco Mountain, Arizona." *Ecology* 46 (1965), 860–64, doi:10.2307/1934019.

Mosher, S., and T. W. Fuller. *Climategate: The Crutape Letters*, vol. 1. N.p.: CreateSpace, 2010.

Munro, M. "Plans to Cool Planet Heat Up Geoengineering Debate." *Vancouver Sun*, 11 May 2010.

National Oceanography Centre, Southampton. "How Well Do Scientists Understand How Changes in Earth's Orbit Affect Long-Term Natural Climate Trends?" ScienceDaily, 7 February 2010, www.sciencedaily.com/releases/2010/02/100205091825.htm, retrieved 7 February 2010.

National Science Foundation. "Shrinking Atmospheric Layer Linked to Low Levels of Solar Radiation." ScienceDaily, 27 August 2010, www.sciencedaily.com/releases/2010/08/100826152217.htm, retrieved 28 August 2010.

Oman, L., et al. "High-Latitude Eruptions Cast Shadow over the African Monsoon and the Flow of the Nile." *Geophysical Research Letters* 33 (2006), L18711, doi:10.1029/2006GL027665.

Oreskes, N. "You Can Argue with the Facts: The Denial of Global Warming." 2008, www.aip.org/history/powerpoints/GlobalWarming_Oreskes.ppt.

Oreskes, N., and E. M. Conway. *Merchants of Doubt: How a Handful of Scientists Obscured the Truth on Issues from Tobacco Smoke to Global Warming*. New York: Bloomsbury, 2010.

Owensby, C. E., R. M. Cochran, and L. M. Auen. "Effects of Elevated Carbon Dioxide on Forage Quality for Ruminants." *Carbon Dioxide, Populations, and Communities*, ed. C. Koerner and F. Bazzaz. Physiologic Ecology Series. San Diego: Academic, 1996.

Owensby, C. E., P. I. Coyne, and L. M. Auen. "Nitrogen and Phosphorus Dynamics of a Tallgrass Prairie Ecosystem Exposed to Elevated Carbon Dioxide." *Plant, Cell and Environment* 16, no. 7 (2006), 843–50.

Paul, F., A. Kääb, M. Maisch, T. Kellenberger, and W. Haeberli. "Rapid Disintegration of Alpine Glaciers Observed with Satellite Data." *Geophysical Research Letters* 31 (2004), L21402, doi:10.1029/2004GL020816.

Pew Center on Global Climate Change. Reports, 2009, www.pewclimate.org/publications, including *Hurricanes and Global Warming FAQs*, www.pewclimate.org/hurricanes, and *Global Climate Change Impacts in the United States: Report of the U.S. Global Change Research Program*, www.pewclimate.org/report/global-climate-change-impacts-in-the-united-states/june-2009.

Pew Research Center for the People and the Press. "Fewer Americans See Solid Evidence of Global Warming." October 2009, people-press.org/report/556/global-warming.

Philander, S. G. *Is the Temperature Rising? The Uncertain Science of Global Warming*. Princeton: Princeton University Press, 1998.

Pielke, Roger. *The Climate Fix: What Scientists and Politicians Won't Tell You about Global Warming*. New York: Basic Books, 2010.

Pilkey, O. H. *A Celebration of the World's Barrier Islands*. New York: Columbia University Press, 2003.

Pilkey, O. H., and L. Pilkey-Jarvis. *Useless Arithmetic: Why Environmental Scientists Can't Predict the Future*. New York: Columbia University Press, 2007.

Pilkey, O. H., and R. Young. *The Rising Sea*. Washington: Island, 2009.

Post, E., et al. "Ecological Dynamics across the Arctic Associated with Recent Climate Change." *Science* 325, no. 5946 (11 September 2009), 1355–58, doi:10.1126/science.1173113.

Pritchard, H. D., et al. "Extensive Dynamic Thinning on the Margins of the Greenland and Antarctic Ice Sheets." *Nature* 461 (2009), 971–75, doi:10.1038/nature08471.

Pugh, D. *Changing Sea Levels: Effects of Tides, Weather and Climate*. Cambridge: Cambridge University Press, 2004.

Real Climate. "Climate Science from Climate Scientists: Calculating the Greenhouse Effect." 21 January 2006, www.realclimate.org/index.php/archives/2006/01/calculating-the-greenhouse-effect.

Reisner, M. *Cadillac Desert: The American West and Its Disappearing Water*. New York: Penguin, 1993.

Revelle, R., and H. Suess. "Carbon Dioxide Exchange between Atmosphere and Ocean and the Question of an Increase of Atmospheric CO_2 during the Past Decades." *Tellus* 9 (1957), 18–27.

Riebesell, U., I. Zondervan, B. Rost, P. D. Tortell, R. E. Zeebe, and F. M. M. Morel. "Reduced Calcification of Marine Plankton in Response to Increased Atmospheric CO_2" (abstract). *Nature* 407, no. 6802 (2000), 364–67, doi:10.1038/35030078.

Romm, J. *Hell and High Water: Global Warming: The Solution and the Politics and What We Should Do*. New York: William Morrow, 2007.

Schneider, S. H. *Science as a Contact Sport: Inside the Battle to Save the Earth's Climate*. Washington: National Geographic, 2009.

Schneider, S. H., A. Rosencranz, M. D. Mastrandrea, and K. Kuntz-Duriseti, eds. *Climate Change Science and Policy*. Washington: Island, 2010.

Sella, G., S. Stein, T. H. Dixon, M. Craymer, T. James, S. Mazzotti, and R. K. Dokka. "Observation of Glacial Isostatic Adjustment in 'stable' North America with GPS." *Geophysical Research Letters* 34 (2006), L02306, doi:10:1-29/2006GL027081.

Shaw, M. R., E. S. Zavaleta, N. R. Chiariello, et al. "Grassland Responses to Global Environmental Changes Suppressed by Elevated CO_2." *Science* 298, no. 5600 (2002), 1987–90, doi:10.1126/science.1075312.

Smith, K., ed. *Nitrous Oxide and Climate Change*. London: Earthscan, 2010.

Society for General Microbiology. "Ecosystems under Threat from Ocean Acidification." ScienceDaily, 31 March 2010, www.sciencedaily.com/releases/2010/03/100329075913.htm, retrieved 26 May 2010.

Solomon, L. *The Deniers, Fully Revised: The World-Renowned Scientists Who Stood Up against Global Warming Hysteria, Political Persecution and Fraud*. Minneapolis: Richard Vigilante, 2010.

Stafford, N. "Future Crops: The Other Greenhouse Effect." *Nature* 448 (2 August 2007), 526–28, doi:10.1038/448526a.

Steffen, A. "Geoengineering and the New Climate Denialism." 29 April 2009, www.worldchanging.com/archives/009784.html.

Stewart, W. *Climate of Uncertainty: A Balanced Look at Global Warming and Renewable Energy*. Flagler Beach, Fla.: Ocean, 2010.

Stirling, I., A. E. Derocher, W. A. Gough, and K. Rode. "Response to Dyck et al. (2007) on Polar Bears and Climate Change in Western

Hudson Bay." *Ecological Complexity* 5, no. 3 (September 2008), 193–201, doi:10.1016/j.ecocom.2008.01.004.

Sussman, B. *Climategate: A Veteran Meteorologist Exposes the Global Warming Scam*. Washington: WND, 2010.

State of the Birds. "Secretary Salazar Releases New 'State of the Birds' Report Showing Climate Change Threatens Hundreds of Species," www.stateofthebirds.org/newsroom/2010-news-release.

Thernstrom, S. *Resetting the Earth's Thermostat*. AEI Outlooks and On the Issues, 27 June 2008, www.aei.org/issue/28202.

Trenberth, K. E., and A. Dai. "Effects of Mount Pinatubo Volcanic Eruption on the Hydrological Cycle as an Analog of Geoengineering." *Geophysical Research Letters* 34 (2007), L15702, doi:10.1029/2007GL030524.

Trumper, K., et al.. "The Natural Fix? The Role of Ecosystems in Climate Fixation." United Nations Environment Programme, 2007.

Turner, J., J. C. Comiso, G. J. Marshall, et al. "Non-annular Atmospheric Circulation Change Induced by Stratospheric Ozone Depletion and Its Role in the Recent Increase of Antarctic Sea Ice Extent." *Geophysical Research Letters* 36 (2009), L08502, doi:10.1029/2009GL037524.

Union of Concerned Scientists. "Smoke, Mirrors, and Hot Air." 2007, www.ucsusa.org/assets/documents/global_warming/exxon_report.pdf.

United Nations Environment Programme. *Climate in Peril: A Popular Guide to the Latest IPCC Reports*. Arendal, Norway: UNEP/GRID-Arendal, 2008.

———. *Climate Change Science Compendium 2009*.

———. *UNEP Year Book 2009*.

———. *UNEP Year Book 2010*.

U.S. Global Change Research Program. *Climate Change Impacts on the United States: The Potential Consequences of Climate Variability and Change*. Cambridge: Cambridge University Press, 2000.

Usoskin, I. G., et al. "Solar Activity over the Last 1,150 Years: Does It Correlate with Climate?" Proceedings of the 13th Cool Stars Workshop, Hamburg, 5–9 July 2004.

Vellinga, P., and R. J. T. Klein. "Climate Change, Sea Level Rise and Integrated Coastal Zone Management: An IPCC Approach." *Ocean and Coastal Management* 21, nos. 1–3 (1993), 245–68.

Wingham, D. J., D. W. Wallis, and A. Sheperd. "The Spatial and Temporal Evolution of Pine Island Glacier Thinning, 1995–2006." *Geophysical Research Letters* 36 (2009), L17501, doi:10.1029/2009GL039126.

WWF Nepal Program. "An Overview of Glaciers, Glacier Retreat and Subsequent Impacts in Nepal, India and China." March 2005.

Zachos, J. C., U. Röhl, S. A. Schellenberg, et al. "Rapid Acidification of the Ocean during the Paleocene-Eocene Thermal Maximum." *Science* 308 (2005), 1611–15.

Zavala, J. A., C. L. Casteel, E. H. DeLucia, and M. R. Berenbaum. 2008: "Anthropogenic Increase in Carbon Dioxide Compromises Plant Defense against Invasive Insects." *Proceedings of the National Academy of Sciences of the United States of America* 105, no. 13 (2008), 5129–33, doi:10.1073/pnas.0800568105 [correction in *Proceedings of the National Academy of Sciences of the United States of America* 105, no. 30 (2008), 10631, doi:10.1073/pnas.0805247105].

Zeman, F. S., and D. W. Keith. "Carbon Neutral Hydrocarbons." *Philosophical Transactions of the Royal Society (A)* 366 (2008), 3901–18.

INDEX

ORRIN H. PILKEY

is the James B. Duke Professor Emeritus of Earth and Ocean Sciences at the Nicholas School of the Environment at Duke University.

KEITH C. PILKEY

is an attorney with a longstanding interest in the role of corporate influence in science and policy.

MARY EDNA FRASER

is a batik artist who highlights environmental concerns and climate change in her art; she employs ancient fabric-dyeing techniques, photography, and conservation science in her work.

Library of Congress Cataloging-in-Publication Data
Pilkey, Orrin H., 1934–
Global climate change : a primer / Orrin H. Pilkey
and Keith C. Pilkey; with batik art by Mary Edna Fraser.
p. cm.
Includes bibliographical references and index.
ISBN 978 0-8223-5095-8 *(cloth : alk. paper)*
ISBN 978 0-8223-5109-2 *(pbk. : alk. paper)*
1. *Climatic changes.* 2. *Global temperature changes.*
3. *Global warming.* I. *Pilkey, Keith C., 1965–*
II. *Fraser, Mary Edna.* III. *Title.*
QD903.P555 2011
363.738'74—*dc22* 2011006493